Quick Review Math Handbook

Book 1

 **Glencoe**

New York, New York   Columbus, Ohio   Chicago, Illinois   Woodland Hills, California

The *McGraw·Hill* Companies

 Glencoe

Send all inquiries to:
Glencoe/McGraw-Hill
8787 Orion Place
Columbus, OH 43240-4027

ISBN: 978-0-07-891504-8   *(Student Edition)*
MHID: 0-07-891504-X   *(Student Edition)*
ISBN: 978-0-07-891505-5   *(Teacher Wraparound Edition)*
MHID: 0-07-891505-8   *(Teacher Wraparound Edition)*

Printed in the United States of America.

1 2 3 4 5 6 7 8 9 10 071 17 16 15 14 13 12 11 10 09 08

Handbook
at a Glance

Handbook Contents

4 Data, Statistics, and Probability 170

5 Algebra 208

6•3 Symmetry and Transformations

6•4 Perimeter

6•5 Area

6•6 Surface Area

6•7 Volume

6•8 Circles

7 Measurement 302

CONTENTS

PART THREE

HotSolutions and Index 350

Handbook
Introduction

Why use this handbook?
You will use this handbook to refresh your memory of mathematics concepts and skills.

What are HotWords, and how do you find them?
HotWords are important mathematical terms. The HotWords section includes a glossary of terms, a collection of common or significant mathematical patterns, and lists of symbols and formulas in alphabetical order. Many entries in the glossary will refer you to chapters and topics in the HotTopics section for more detailed information.

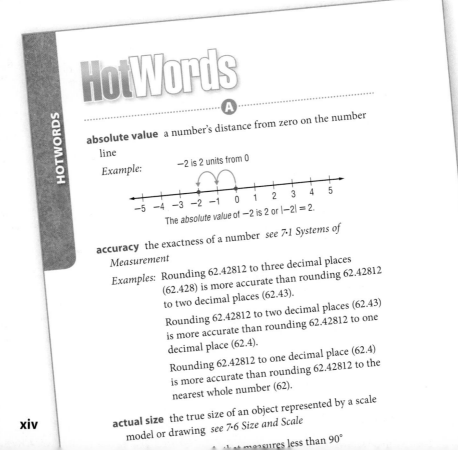

HOTWORDS

HotWords

A

absolute value a number's distance from zero on the number line

Example: −2 is 2 units from 0

The *absolute value* of −2 is 2 or |−2| = 2.

accuracy the exactness of a number *see 7·1 Systems of Measurement*

Examples: Rounding 62.42812 to three decimal places (62.428) is more accurate than rounding 62.42812 to two decimal places (62.43).

Rounding 62.42812 to two decimal places (62.43) is more accurate than rounding 62.42812 to one decimal place (62.4).

Rounding 62.42812 to one decimal place (62.4) is more accurate than rounding 62.42812 to the nearest whole number (62).

actual size the true size of an object represented by a scale model or drawing *see 7·6 Size and Scale*

What are HotTopics, and how do you use them?

HotTopics are key concepts that you need to know. The HotTopics section consists of eight chapters. Each chapter has several topics that give you to-the-point explanations of key mathematical concepts. Each topic includes one or more concepts. Each section includes Check It Out exercises, which you can use to check your understanding. At the end of each topic, there is an exercise set.

There are problems and a vocabulary list at the beginning and end of each chapter to help you preview what you know and review what you have learned.

What are HotSolutions?

The HotSolutions section gives you easy-to-locate answers to Check It Out and What Do You Know? problems. The HotSolutions section is at the back of the handbook.

HotSolutions

Chapter ❶ Numbers and C

p. 66

1. 30,000   2. 30,000,000
3. $(2 \times 10,000) + (4 \times 1,000) + (3$
   $+ (7 \times 10) + (8 \times 1)$
4. 566,418; 496,418; 56,418; 5,618
5. 52,564,760; 52,565,000; 53,000,
7. 15   8. 400   9. 1,600   10. (4 +
11. $(10 + 14) \div (3 + 3) = 4$ 12.
14. no   15. yes   16. 3×11 17.
18. $2 \times 2 \times 3 \times 3 \times 5$

p. 67

19. 15   20. 7   21. 6   22. 15   2
25. 6   26. −13   27. 15   28. −2
31. −18   32. 0   33. 28   34. −4
37. −30   38. −60   39. It will b
40. It will be a positive integer.

1•1 Place Value of Whole N

1•3 Order of Op

Understanding the Order of Op

Solving a problem may involve using more th
Your answer can depend on the order in whic
operations.

For instance, consider the expression $2 + 3 \times$

$$2 + 3 \times 4$$
$$5 \times 4 = 20$$

or

$$2 + 3 \times 4$$
$$2 + 12 = 14$$

The order in which you perform operations makes a differe

To make sure that there is just one answer to a se
computations, mathematicians have agreed on an
which to do the operations.

EXAMPLE Using the Order of Operations

Simplify $2 + 8 \times (9 - 5)$.
$2 + 8 \times (9 - 5)$
   $= 2 + 8 \times 4$
$2 + 8 \times 4 = 2 + 32$
$2 + 32 = 34$
So, $2 + 8 \times (9 - 5) = 34$.

• Simplify within the parenti
  any powers. (See p. 158.)
• Multiply or divide from left
• Add or subtract from left to

Check It Out
Simplify.

❶ $20 - 2 \times 5$

❸ $(8 \times 2) - 16$

❷ $3 \times (2 + 16$

Hot Words

The **HotWords** section includes a glossary of terms, lists of formulas and symbols, and a collection of common or significant mathematical patterns. Many entries in the glossary will refer to chapters and topics in the **HotTopics** section.

HotWords

A

absolute value a number's distance from zero on the number line

Example: −2 is 2 units from 0

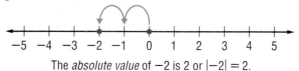

The *absolute value* of −2 is 2 or |−2| = 2.

accuracy the exactness of a number *see 7·1 Systems of Measurement*

Examples: Rounding 62.42812 to three decimal places (62.428) is more accurate than rounding 62.42812 to two decimal places (62.43).

Rounding 62.42812 to two decimal places (62.43) is more accurate than rounding 62.42812 to one decimal place (62.4).

Rounding 62.42812 to one decimal place (62.4) is more accurate than rounding 62.42812 to the nearest whole number (62).

actual size the true size of an object represented by a scale model or drawing *see 7·6 Size and Scale*

acute angle any angle that measures less than 90°
see 6·1 Naming and Classifying Angles and Triangles

Example:

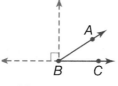

∠*ABC* is an *acute angle*.
0° < *m∠ABC* < 90°

acute triangle a triangle in which all angles measure less than 90° *see 6·1 Naming and Classifying Angles and Triangles*

Example:

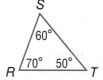

△*RST* is an *acute triangle.*

Addition Property of Equality the mathematical rule that states that if the same number is added to each side of an equation, the expressions remain equal

Example: If $a = b$, then $a + c = b + c$.

additive inverse two integers that are opposite of each other; the sum of any number and its *additive inverse* is zero

Example: $(+3) + (-3) = 0$
(-3) is the *additive inverse* of 3.

additive system a mathematical system in which the values of individual symbols are added together to determine the value of a sequence of symbols

Example: The Roman numeral system, which uses symbols such as I, V, D, and M, is a well-known *additive system.*

This is another example of an additive system:
    ▽▽□
    If □ equals 1 and ▽ equals 7,
    then ▽▽□ equals $7 + 7 + 1 = 15$.

algebra a branch of mathematics in which symbols are used to represent numbers and express mathematical relationships *see Chapter 5 Algebra*

algorithm a step-by-step procedure for a mathematical operation

altitude the perpendicular distance from a vertex to the opposite side of a figure; *altitude* indicates the height of a figure

Example:

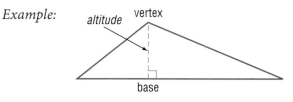

angle two rays that meet at a common endpoint *see 6·1 Naming and Classifying Angles and Triangles, 8·3 Geometry Tools*

Example:

∠*ABC* is formed by $\overrightarrow{BA}$ and $\overrightarrow{BC}$.

angle of elevation the angle formed by a horizontal line and an upward line of sight

Example:

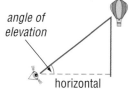

apothem a perpendicular line segment from the center of a regular polygon to one of its sides

Example:

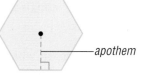

approximation an estimate of a mathematical value *see 1·1 Place Value of Whole Numbers*

Arabic numerals (or Hindu-Arabic numerals) the number symbols we presently use in our base-ten number system {0, 1, 2, 3, 4, 5, 6, 7, 8, 9}

arc a section of a circle

Example:

$\overset{\frown}{QR}$ is an *arc*.

area the measure of the interior region of a 2-dimensional figure or the surface of a 3-dimensional figure, expressed in square units *see Formulas page 58, 3·1 Powers and Exponents, 6·5 Area, 6·6 Surface Area, 6·8 Circles, 7·1 Systems of Measurements, 7·3 Area, Volume, and Capacity*

Example:

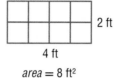

4 ft

2 ft

area = 8 ft²

arithmetic expression a mathematical relationship expressed as a number, or two or more numbers with operation symbols *see expression*

arithmetic sequence *see Patterns page 61, 6·7 Graphing on the Coordinate Plane*

Associative Property the mathematical rule that states that the way in which numbers are grouped when they are added or multiplied does not change their sum or product *see 5·2 Simplifying Expressions*

Examples: $(x + y) + z = x + (y + z)$
$x \times (y \times z) = (x \times y) \times z$

average the sum of a set of values divided by the number of values *see 4·3 Statistics*

Example: The *average* of 3, 4, 7, and 10 is
$(3 + 4 + 7 + 10) \div 4 = 6.$

average speed the average rate at which an object moves

axis (pl. *axes*) [1] a reference line by which a point on a coordinate graph may be located; [2] the imaginary line about which an object may be said to be symmetrical (*axis* of symmetry); [3] the line about which an object may revolve (*axis* of rotation) *see 5·6 Graphing on the Coordinate Plane, 6·3 Symmetry and Transformations*

· **B** ·

bar graph a display of data that uses horizontal or vertical bars to compare quantities *see 4·2 Displaying Data*

base [1] the number used as the factor in exponential form; [2] two parallel congruent faces of a prism or the face opposite the apex of a pyramid or cone; [3] the side perpendicular to the height of a polygon; [4] the number of characters in a number system *see 1·1 Place Value of Whole Numbers, 3·1 Powers and Exponents, 6·5 Area*

base-ten system the number system containing ten single-digit symbols {0, 1, 2, 3, 4, 5, 6, 7, 8, and 9} in which the numeral 10 represents the quantity ten *see 2·5 Naming and Ordering Decimals*

base-two system the number system containing two single-digit symbols {0 and 1} in which 10 represents the quantity two *see binary system*

benchmark a point of reference from which measurements and percents can be estimated *see 2·7 Meaning of Percent*

best chance in a set of values, the event most likely to occur *see 4·4 Probability*

binary system the base-two number system, in which combinations of the digits 1 and 0 represent different numbers or values

binomial an algebraic expression that has two terms

Examples: $x^2 + y; x + 1; a - 2b$

budget a spending plan based on an estimate of income and expenses

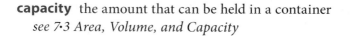

capacity the amount that can be held in a container
see 7·3 Area, Volume, and Capacity

cell a small rectangle in a spreadsheet that stores information; each *cell* can store a label, number, or formula
see 8·4 Spreadsheets

center of the circle the point from which all points on a circle are equidistant *see 6·8 Circles*

chance the probability or likelihood of an occurrence, often expressed as a fraction, decimal, percentage, or ratio
see 2·9 Fraction, Decimal, and Percent Relationships,
4·4 Probability, 5·5 Ratio and Proportion

circle the set of all points in a plane that are equidistant from a fixed point called the center *see 6·8 Circles, 8·1 Four-Function Calculator*

Example:

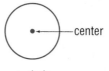

center

a *circle*

circle graph (pie chart) a display of statistical data that uses a circle divided into proportionally-sized "slices"
see 4·2 Displaying Data

Example:

Favorite Primary Color

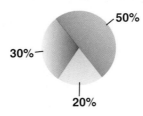

50%

30%—

20%

circumference the distance around (perimeter) a circle
see Formulas page 59, 6·6 Circles

classification the grouping of elements into separate classes or sets

collinear a set of points that lie on the same line

Example:

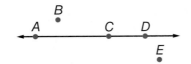

Points *A, C,* and *D* are *collinear.*

columns vertical lists of numbers or terms *see 8·4 Spreadsheets*

combination a selection of elements from a larger set in which the order does not matter

Example: 456, 564, and 654 are one *combination* of three digits from 4567.

common denominator a common multiple of the denominators of a group of fractions *see 2·2 Addition and Subtraction of Fractions*

Example: The fractions $\frac{3}{4}$ and $\frac{7}{8}$ have a *common denominator* of 8.

common difference the difference between any two consecutive terms in an arithmetic sequence

common factor a whole number that is a factor of each number in a set of numbers *see 1·4 Factors and Multiples*

Example: 5 is a *common factor* of 10, 15, 25, and 100.

common ratio the ratio of any term in a geometric sequence to the term that precedes it

Commutative Property the mathematical rule that states that the order in which numbers are added or multiplied does not change their sum or product *see 5·2 Simplifying Expressions*

Examples: $x + y = y + x$
$x \times y = y \times x$

compatible numbers two numbers that are easy to add, subtract, multiply, or divide mentally

complementary angles two angles are complementary if the sum of their measures is 90° *see 6·1 Classifying Angles and Triangles*

∠1 and ∠2 are complementary angles.

composite number a whole number greater than 1 having more than two factors *see 1·4 Factors and Multiples*

concave polygon a polygon that has an interior angle greater than 180°

Example:

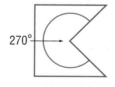

a *concave polygon*

cone a three-dimensional figure consisting of a circular base
and one vertex

Example:

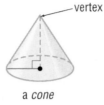

a *cone*

congruent having the same size and shape; the symbol ≅ is
used to indicate congruence *see 6·1 Classifying Angles and
Triangles*

Example:

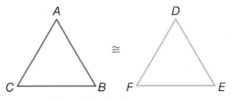

△*ABC* and △*DEF* are *congruent.*

congruent angles angles that have the same measure
see 6·1 Naming and Classifying Angles and Triangles

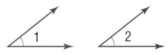

∠1 and ∠2 are congruent angles.

conic section the curved shape that results when a conical
surface is intersected by a plane

Example:

This ellipse
is a *conic section.*

continuous data the complete range of values on the
number line

Example: The possible sizes of apples are *continuous data.*

convex polygon a polygon with all interior angles measuring less than 180°

Example:

A regular hexagon is a *convex polygon.*

coordinate any number within a set of numbers that is used to define a point's location on a line, on a surface, or in space *see 5·7 Graphing on a Coordinate Plane*

Example:

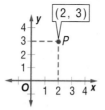

Point *P* has *coordinates* (2, 3).

coordinate plane a plane in which a horizontal number line and a vertical number line intersect at their zero points *see 5·7 Graphing on the Coordinate Plane*

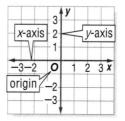

coplanar points or lines lying in the same plane

correlation the way in which a change in one variable corresponds to a change in another

cost an amount paid or required in payment

cost estimate an approximate amount to be paid or to be required in payment

counting numbers the set of positive whole numbers
{1, 2, 3, 4 . . .} *see positive integers*

cross product a method used to solve proportions and test
whether ratios are equal *see 2·1 Fractions and Equivalent
Fractions, 5·5 Ratio and Proportion*

Example: $\frac{a}{b} = \frac{c}{d}$ if $a \times d = b \times c$

cross section the figure formed by the intersection of a solid
and a plane

Example:

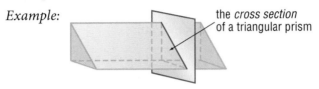

the *cross section*
of a triangular prism

cube [1] a solid figure with six congruent square faces
see 6·2 Polygons and Polyhedrons [2] the product of three
equal terms *see 3·1 Powers and Exponents, 8·2 Scientific
Calculator*

Examples: [1]

a *cube*

[2] $2^3 = 2 \times 2 \times 2 = 8$

cube root the number that when raised to the third power
equals a given number *see 8·2 Scientific Calculator*

Example: $\sqrt[3]{8} = 2$

2 is the *cube root* of 8.

cubic centimeter the volume of a cube with edges that are 1 centimeter in length *see 6·7 Volume*

cubic foot the volume of a cube with edges that are 1 foot in length *see 6·7 Volume*

cubic inch the volume of a cube with edges that are 1 inch in length *see 6·7 Volume*

cubic meter the volume of a cube with edges that are 1 meter in length *see 6·7 Volume*

customary system units of measurement used in the United States to measure length in inches, feet, yards, and miles; capacity in cups, pints, quarts, and gallons; weight in ounces, pounds, and tons; and temperature in degrees Fahrenheit *see English system, 7·1 Systems of Measurement*

cylinder a solid shape with parallel circular bases

Example:

a *cylinder*

·································· **D** ··································

decagon a polygon with ten angles and ten sides *see 6·2 Polygons and Polyhedrons*

decimal system the most commonly used number system, in which whole numbers and fractions are represented using base ten *see 2·5 Naming and Ordering Decimals, 8·2 Scientific Calculator*

Example: Decimal numbers include 1230, 1.23, 0.23, and −13.

degree [1] (algebraic) the exponent of a single variable in a simple algebraic term; [2] (algebraic) the sum of the exponents of all the variables in a more complex algebraic term; [3] (algebraic) the highest degree of any term in a polynomial; [4] (geometric) a unit of measurement of an angle or arc, represented by the symbol ° *see 3·1 Powers and Exponents, 6·1 Naming and Classifying Angles and Triangles, 6·6 Circles, 8·2 Scientific Calculator*

Examples: [1] In the term $2x^4y^3z^2$, x has a *degree* of 4, y has a *degree* of 3, and z has a *degree* of 2.

[2] The term $2x^4y^3z^2$ as a whole has a *degree* of $4 + 3 + 2 = 9$.

[3] The equation $x^3 = 3x^2 + x$ is an equation of the third *degree*.

[4] An acute angle is an angle that measures less than 90°.

denominator the bottom number in a fraction representing the total number of equal parts in the whole *see 2·1 Fractions and Equivalent Fractions*

Example: In the fraction $\frac{a}{b}$, b is the *denominator*.

dependent events two events in which the outcome of one event is affected by the outcome of another event

diagonal a line segment connecting two non-adjacent vertices of a polygon *see 6·2 Polygons and Polyhedrons*

Example:

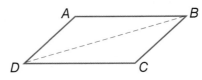

$\overline{BD}$ is a *diagonal* of parallelogram *ABCD*.

diameter a line segment connecting the center of a circle with two points on its perimeter *see 6·8 Circles*

Example:

diameter

difference the result obtained when one number is subtracted from another *see 5·1 Writing Expressions and Equations*

dimension the number of measures needed to describe a figure geometrically

Examples: A point has 0 *dimensions.*
A line or curve has 1 *dimension.*
A plane figure has 2 *dimensions.*
A solid figure has 3 *dimensions.*

direct correlation the relationship between two or more elements that increase and decrease together

Example: At an hourly pay rate, an increase in the number of hours worked means an increase in the amount paid, while a decrease in the number of hours worked means a decrease in the amount paid.

discount a deduction made from the regular price of a product or service

discrete data only a finite number of values is possible

Example: The number of parts damaged in a shipment is *discrete data.*

distance the length of the shortest line segment between two points, lines, planes, and so forth *see 7·2 Length and Distance, 8·3 Geometry Tools*

Distributive Property the mathematical rule that states that multiplying a sum by a number gives the same result as multiplying each addend by the number and then adding the products *see 1·2 Properties, 5·2 Simplifying Expressions*

Example: $a(b + c) = a \times b + a \times c$

divisible a number is *divisible* by another number if their quotient has no remainder

division the operation in which a dividend is divided by a divisor to obtain a quotient

Example:

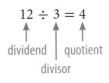

$$12 \div 3 = 4$$

dividend | quotient
divisor

Division Property of Equality the mathematical rule that states that if each side of an equation is divided by the same nonzero number, the two sides remain equal

Example: If $a = b$, then $\frac{a}{c} = \frac{b}{c}$.

domain the set of input values in a function

double-bar graph a display of data that uses paired horizontal or vertical bars to compare quantities

Example:

edge a line segment joining two planes of a polyhedron
see 6·2 Polygons and Polyhedrons

English system units of measurement used in the United States
that measure length in inches, feet, yards, and miles; capacity
in cups, pints, quarts, and gallons; weight in ounces, pounds,
and tons; and temperature in degrees Fahrenheit
see customary system

equal angles angles that measure the same number of
degrees *see 7·1 Naming and Classifying Angles and Triangles*

equally likely describes outcomes or events that have the same
chance of occurring *see 4·4 Probability*

equally unlikely describes outcomes or events that have the
same chance of not occurring *see 4·4 Probability*

equation a mathematical sentence stating that two expressions
are equal *see 5·1 Writing Expressions and Equations,
5·4 Equations*

Example: $3 \times (7 + 8) = 9 \times 5$

equiangular the property of a polygon in which all angles are
congruent

equiangular triangle a triangle in which each angle is 60°
see 6·1 Naming and Classifying Angles and Triangles

Example:

$m\angle A = m\angle B = m\angle C = 60°$
$\triangle ABC$ is *equiangular.*

equilateral the property of a polygon in which all sides are congruent

equilateral triangle a triangle in which all sides are congruent

Example:

$AB = BC = AC$
$\triangle ABC$ is *equilateral.*

equivalent equal in value *see 2·1 Fractions and Equivalent Fractions, 5·4 Equations*

equivalent expressions expressions that always result in the same number, or have the same mathematical meaning for all replacement values of their variables *see 5·2 Simplifying Expressions*

Examples: $\frac{9}{3} + 2 = 10 - 5$
$2x + 3x = 5x$

equivalent fractions fractions that represent the same quotient but have different numerators and denominators *see 2·1 Fractions and Equivalent Fractions*

Example: $\frac{5}{6} = \frac{15}{18}$

equivalent ratios ratios that are equal *see 5·4 Ratio and Proportion*

Example: $\frac{5}{4} = \frac{10}{8}; 5:4 = 10:8$

estimate an approximation or rough calculation *see 2·3 Addition and Subtraction of Fractions*

even number any whole number that is a multiple of 2 {0, 2, 4, 6, 8, 10, 12 . . .}

event any happening to which probabilities can be assigned *see 4·4 Probability*

expanded form a method of writing a number that highlights the value of each digit *see 1·1 Place Value of Whole Numbers, 2·1 Fractions and Equivalent Fractions*

Example: $867 = (8 \times 100) + (6 \times 10) + (7 \times 1)$

expense an amount of money paid; cost

experimental probability the ratio of the total number of times the favorable outcome occurs to the total number of times the experiment is completed *see 4·4 Probability*

exponent a numeral that indicates how many times a number or variable is used as a factor *see 3·1 Powers and Exponents*

Example: In the equation $2^3 = 8$, the *exponent* is 3.

expression a mathematical combination of numbers, variables, and operations *see 5·1 Writing Expressions and Equations, 5·2 Simplifying Expressions, 5·3 Evaluating Expressions and Formulas*

Example: $6x + y^2$

································**F**································

face a two-dimensional side of a three-dimensional figure *see 6·2 Polygons and Polyhedrons, 6·6 Surface Area*

factor a number or expression that is multiplied by another to yield a product *see 1·4 Factors and Multiples, 2·4 Multiplication and Division of Fractions, 3·1 Powers and Exponents, 6·1 Naming and Classifying Angles and Triangles*

Example: 3 and 11 are *factors* of 33

factorial represented by the symbol !, the product of all the whole numbers between 1 and a given positive whole number

Example: $5! = 1 \times 2 \times 3 \times 4 \times 5 = 120$

factor pair two unique numbers multiplied together to yield a product

fair describes a situation in which the theoretical probability of each outcome is equal

Fibonacci numbers *see Patterns page 61*

flip a transformation that produces the mirror image of a figure
see 6·3 Symmetry and Transformations

Example:

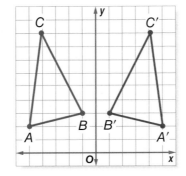

△*A′B′C′* is a *flip* of △*ABC*.

formula an equation that shows the relationship between two or more quantities; a calculation performed by a spreadsheet
see Formulas pages 58–59, 5·3 Evaluating Expressions and Formulas, 8·3 Spreadsheets

Example: $A = \pi r^2$ is the *formula* for calculating the area of a circle; A2 × B2 is a spreadsheet *formula*.

fraction a number representing part of a whole; a quotient in the form $\frac{a}{b}$ *see 2·1 Fractions and Equivalent Fractions*

function the assignment of exactly one output value to each input value *see 5·4 Equations*

Example: You are driving at 50 miles per hour. There is a relationship between the amount of time you drive and the distance you will travel. You say that the distance is a *function* of the time.

geometric sequence *see Patterns page 61*

geometry the branch of mathematics that investigates the relations, properties, and measurement of solids, surfaces, lines, and angles *see Chapter 6 Geometry, 8·3 Geometry Tools*

gram a metric unit of mass *see 7·3 Systems of Measurement*

greatest common factor (GCF) the greatest number that is a factor of two or more numbers *see 1·4 Factors and Multiples, 2·1 Fractions and Equivalent Fractions*

Example: 30, 60, 75

The *greatest common factor* is 15.

harmonic sequence *see Patterns page 61*

height the perpendicular distance from a vertex to the opposite side of a figure *see 6·5 Area, 6·7 Volume*

heptagon a polygon with seven angles and seven sides *see 6·2 Polygons and Polyhedrons*

Example:

a *heptagon*

hexagon a polygon with six angles and six sides *see 6·2 Polygons and Polyhedrons*

Example:

a *hexagon*

hexagonal prism a prism that has two hexagonal bases and six rectangular sides *see 6·2 Polygons and Polyhedrons*

Example:

a *hexagonal prism*

hexahedron a polyhedron that has six faces *see 6·2 Polygons and Polyhedrons*

Example:

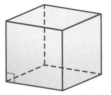

A cube is a *hexahedron.*

horizontal parallel to or in the plane of the horizon *see 5·7 Graphing on the Coordinate Plane, 8·4 Spreadsheets*

hypotenuse the side opposite the right angle in a right triangle *see 6·1 Naming and Classifying Angles and Triangles*

Example:

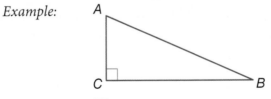

Side $\overline{AB}$ is the *hypotenuse* of this right triangle.

··· **I** ···

improper fraction a fraction in which the numerator is greater than the denominator *see 2·1 Fractions and Equivalent Fractions*

Examples: $\dfrac{21}{4}, \dfrac{4}{3}, \dfrac{2}{1}$

income the amount of money received for labor, services, or the sale of goods or property

independent event two events in which the outcome of one event is not affected by the outcome of another event

inequality a statement that uses the symbols $>$ (greater than), $<$ (less than), $\geq$ (greater than or equal to), and $\leq$ (less than or equal to) to compare quantities *see 5·6 Inequalities*

Examples: $5 > 3; \frac{4}{5} < \frac{5}{4}; 2(5 - x) > 3 + 1$

infinite, nonrepeating decimal irrational numbers, such as π and $\sqrt{2}$, that are decimals with digits that continue indefinitely but do not repeat

inscribed figure a figure that is enclosed by another figure as shown below

Examples:

a triangle *inscribed* in a circle

a circle *inscribed* in a triangle

integers the set of all whole numbers and their additive inverses $\{\ldots, -5, -4, -3, -2, -1, 0, 1, 2, 3, 4, 5, \ldots\}$ *see 1·5 Integer Operations*

intercept [1] the cutting of a line, curve, or surface by another line, curve, or surface; [2] the point at which a line or curve cuts across a coordinate axis

intersection the set of elements common to two or more sets *see Venn diagram, 1·4 Factors and Multiples*

Example:

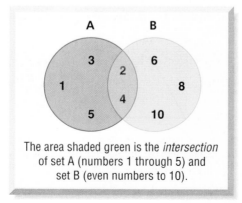
The area shaded green is the *intersection* of set A (numbers 1 through 5) and set B (even numbers to 10).

inverse operation the operation that reverses the effect of another operation *see 2·4 Multiplication and Division of Fractions*

Examples: Subtraction is the *inverse operation* of addition.
Division is the *inverse operation* of multiplication.

irrational numbers the set of all numbers that cannot be expressed as finite or repeating decimals

Example: $\sqrt{2}$ (1.414214 . . .) and π (3.141592 . . .) are *irrational numbers*.

isometric drawing a two-dimensional representation of a three-dimensional object in which parallel edges are drawn as parallel lines *see two-dimensional, three-dimensional*

Example:

isosceles trapezoid a trapezoid in which the two nonparallel sides are of equal length

Example:

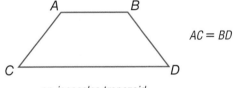

$AC = BD$

an *isosceles trapezoid*

isosceles triangle a triangle with at least two sides of equal length *see 6·1 Naming and Classifying Angles and Triangles*

Example:

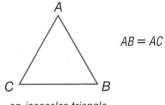

$AB = AC$

an *isosceles triangle*

leaf the unit digit of an item of numerical data between 1 and 99
see stem-and-leaf plot, 4·2 Displaying Data

least common denominator (LCD) the least common multiple
of the denominators of two or more fractions *see 2·2 Addition
and Subtraction of Fractions*

Example: The *least common denominator* of $\frac{1}{3}$, $\frac{2}{4}$, and $\frac{3}{6}$ is 12.

least common multiple (LCM) the smallest nonzero whole
number that is a multiple of two or more whole numbers
*see 1·4 Factors and Multiples, 2·3 Addition and Subtraction of
Fractions*

Example: The *least common multiple* of 3, 9, and 12 is 36.

legs of a triangle the sides adjacent to the right angle of a right
triangle

Example:

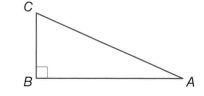

$\overline{AB}$ and $\overline{BC}$ are the *legs of* △*ABC*.

length a measure of the distance of an object from end to end
see 7·2 Length and Distance

likelihood the chance of a particular outcome occurring
see 4·4 Probability

like terms terms that include the same variables raised to the
same powers; *like terms* can be combined *see 5·2 Simplifying
Expressions*

Example: $5x^2$ and $6x^2$ are *like terms;* $3xy$ and $3zy$ are not
like terms.

line a connected set of points extending forever in both
directions *see 5·1 Graphing on a Coordinate Plane,
6·1 Naming and Classifying Angles and Triangles*

linear measure the measure of the distance between two points on a line

line graph a display of data that shows change over time
see 4·2 Displaying Data

Example:

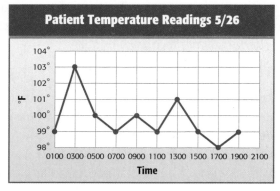

line graph

line of symmetry a line along which a figure can be folded so that the two resulting halves match *see 6·3 Symmetry and Transformations*

Example:

$\overline{ST}$ is a *line of symmetry.*

line plot a display of data that shows the frequency of data on a number line *see 4·2 Displaying Data*

line segment a section of a line between two points
see 6·1 Naming and Classifying Angles and Triangles

Example: A •————————• B

$\overline{AB}$ is a *line segment.*

liter a metric unit of capacity *see 7·3 Area, Volume, and Capacity*

lowest common multiple the smallest number that is a multiple of all the numbers in a given set; same as least common multiple *see 1·4 Factors and Multiples*

Example: The *lowest common multiple* of 6, 9, and 18 is 18.

Lucas numbers *see Patterns page 61*

....**M**..

magic square *see Patterns page 62*

maximum value the greatest value of a function or a set of numbers

mean the quotient obtained when the sum of the numbers in a set is divided by the number of addends *see average, 4·3 Statistics*

Example: The *mean* of 3, 4, 7, and 10 is
$$(3 + 4 + 7 + 10) \div 4 = 6.$$

measurement units standard measures, such as the meter, the liter, and the gram, or the foot, the quart, and the pound *see 7·1 Systems of Measurement*

median the middle number in an ordered set of numbers *see 4·3 Statistics*

Example: 1, 3, 9, 16, 22, 25, 27
16 is the *median*.

meter the metric unit of length

metric system a decimal system of weights and measurements based on the meter as its unit of length, the kilogram as its unit of mass, and the liter as its unit of capacity *see 7·1 Systems of Measurement*

midpoint the point on a line segment that divides it into two equal segments

Example:

$$A \bullet\!\!\!\!\xrightarrow{\hspace{1.2cm}M\hspace{1.2cm}}\!\!\!\!\bullet B$$

$$AM = MB$$

M is the *midpoint* of $\overline{AB}$.

minimum value the least value of a function or a set of numbers

mixed number a number composed of a whole number and a fraction *see 2·1 Fractions and Equivalent Fractions*

Example: $5\frac{1}{4}$ is a *mixed number.*

mode the number or element that occurs most frequently in a set of data *see 4·3 Statistics*

Example: 1, 1, 2, 2, 3, 5, 5, 6, 6, 6, 8
6 is the *mode.*

monomial an algebraic expression consisting of a single term

Example: $5x^3y$, *xy*, and *2y* are three *monomials.*

multiple the product of a given number and an integer
see 1·4 Factors and Multiples

Examples: 8 is a *multiple* of 4.
3.6 is a *multiple* of 1.2.

multiplication one of the four basic arithmetical operations, involving the repeated addition of numbers

multiplication growth number a number that when used to multiply a given number a given number of times results in a given goal number

Example: Grow 10 into 40 in two steps by multiplying
$(10 \times 2 \times 2 = 40)$
2 is the *multiplication growth number.*

Multiplication Property of Equality the mathematical rule that states that if each side of an equation is multiplied by the same number, the two sides remain equal

Example: If $a = b$, then $a \times c = b \times c$.

multiplicative inverse two numbers are multiplicative inverses if their product is 1 *see 2·4 Multiplication and Division of Fractions*

Example: $10 \times \dfrac{1}{10} = 1$

$\dfrac{1}{10}$ is the *multiplicative inverse* of 10.

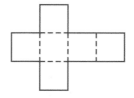

negative integers the set of all integers that are less than zero $\{-1, -2, -3, -4, -5, \ldots\}$ *see 1·5 Integer Operations*

negative numbers the set of all real numbers that are less than zero $\{-1, -1.36, -\sqrt{2}, -\pi\}$ *see 1·5 Integer Operations, 8·1 Four-Function Calculator*

net a two-dimensional plan that can be folded to make a three-dimensional model of a solid *see 6·6 Surface Area*

Example:

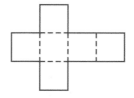

the *net* of a cube

nonagon a polygon with nine angles and nine sides *see 6·2 Polygons and Polyhedrons*

Example:

a *nonagon*

noncollinear points not lying on the same line

noncoplanar points or lines not lying on the same plane

number line a line showing numbers at regular intervals on which any real number can be indicated *see 5·6 Inequalities*

Example:

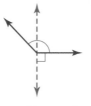

a *number line*

number symbols the symbols used in counting and measuring

Examples: $1, -\frac{1}{4}, 5, \sqrt{2}, -\pi$

number system a method of writing numbers; the Arabic *number system* is most commonly used today *see 1·1 Place Value of Whole Numbers*

numerator the top number in a fraction representing the number of equal parts being considered *see 2·1 Fractions and Equivalent Fractions*

Example: In the fraction $\frac{a}{b}$, a is the *numerator.*

.... **O** ....

obtuse angle any angle that measures greater than 90° but less than 180° *see 6·1 Naming and Classifying Angles and Triangles*

Example:

an *obtuse angle*

obtuse triangle a triangle that has one obtuse angle *see 6·1 Naming and Classifying Angles and Triangles*

Example:

△*ABC* is an *obtuse triangle.*

octagon a polygon with eight angles and eight sides
see *6·2 Polygons and Polyhedrons*

Example:

an *octagon*

octagonal prism a prism that has two octagonal bases and
eight rectangular faces *see 6·2 Polygons and Polyhedrons*

Example:

an *octagonal prism*

odd numbers the set of all integers that are not multiples of 2

odds against the ratio of the number of unfavorable outcomes
to the number of favorable outcomes *see 4·4 Probability*

odds for the ratio of the number of favorable outcomes to the
number of unfavorable outcomes *see 4·4 Probability*

one-dimensional having only one measurable quality
see *Chapter 6 Geometry*

Example: A line and a curve are *one-dimensional.*

operations arithmetical actions performed on numbers,
matrices, or vectors *see 1·3 Order of Operations*

opposite angle in a triangle, a side and an angle are said to be
opposite if the side is not used to form the angle
see *6·2 Polygons and Polyhedrons*

Example:

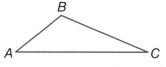

In △*ABC*, ∠*A* is opposite of $\overline{BC}$.

ordered pair two numbers that tell the x-coordinate and y-coordinate of a point *see 5·7 Graphing on the Coordinate Plane*

Example: The coordinates (3, 4) are an *ordered pair*. The x-coordinate is 3, and the y-coordinate is 4.

order of operations to simplify an expression, follow this four-step process: 1) do all operations within parentheses; 2) simplify all numbers with exponents; 3) multiply and divide in order from left to right; 4) add and subtract in order from left to right *see 1·1 Order of Operations, 5·3 Evaluating Expressions and Formulas*

origin the point (0, 0) on a coordinate graph where the x-axis and the y-axis intersect *see 5·7 Graphing on the Coordinate Plane*

outcome a possible result in a probability experiment *see 4·5 Combinations and Permutations, 4·4 Probability*

outcome grid a visual model for analyzing and representing theoretical probabilities that shows all the possible outcomes of two independent events *see 4·4 Probability*

Example:

A grid used to find the sample space for rolling a pair of dice. The outcomes are written as ordered pairs.

	1	2	3	4	5	6
1	(1, 1)	(2, 1)	(3, 1)	(4, 1)	(5, 1)	(6, 1)
2	(1, 2)	(2, 2)	(3, 2)	(4, 2)	(5, 2)	(6, 2)
3	(1, 3)	(2, 3)	(3, 3)	(4, 3)	(5, 3)	(6, 3)
4	(1, 4)	(2, 4)	(3, 4)	(4, 4)	(5, 4)	(6, 4)
5	(1, 5)	(2, 5)	(3, 5)	(4, 5)	(5, 5)	(6, 5)
6	(1, 6)	(2, 6)	(3, 6)	(4, 6)	(5, 6)	(6, 6)

There are 36 possible outcomes.

outlier data that are more than 1.5 times the interquartile range from the upper or lower quartiles *see 4·3 Statistics*

parallel straight lines or planes that remain a constant distance from each other and never intersect, represented by the symbol ‖ *see 6·2 Polygons and Polyhedrons*

Example:

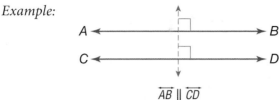

$$\overleftrightarrow{AB} \parallel \overleftrightarrow{CD}$$

parallelogram a quadrilateral with two pairs of parallel sides *see 6·2 Polygons and Polyhedrons*

Example:

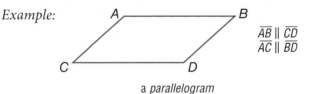

$$\overline{AB} \parallel \overline{CD}$$
$$\overline{AC} \parallel \overline{BD}$$

a *parallelogram*

parentheses the enclosing symbols ( ), which indicate that the terms within are a unit *see 8·2 Scientific Calculator*

Example: $(2 + 4) \div 2 = 3$

Pascal's Triangle *see Patterns page 62*

pattern a regular, repeating design or sequence of shapes or numbers *see Patterns pages 61–63*

PEMDAS an acronym for the order of operations: 1) do all operations within **p**arentheses; 2) simplify all numbers with **e**xponents; 3) **m**ultiply and **d**ivide in order from left to right; 4) **a**dd and **s**ubtract in order from left to right *see 1·3 Order of Operations*

pentagon a polygon with five angles and five sides
see 6·2 Polygons and Polyhedrons

Example:

a *pentagon*

percent a number expressed in relation to 100, represented by
the symbol % *see 2·7 Meaning of Percents, 4·2 Displaying
Data, 8·1 Four-Function Calculator*

Example: 76 out of 100 students use computers.
76 *percent* or 76% of students use computers.

percent grade the ratio of the rise to the run of a hill, ramp, or
incline expressed as a percent

Example:

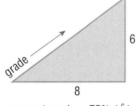

percent grade = 75% ($\frac{6}{8}$)

percent proportion compares part of a quantity to the whole
quantity using a percent *see 2·7 Using and Finding Percents*

$$\frac{part}{whole} = \frac{percent}{100}$$

perfect cube a number that is the cube of an integer

Example: 27 is a *perfect cube* since $27 = 3^3$.

perfect number an integer that is equal to the sum of all its
positive whole number divisors, excluding the number itself

Example: $1 \times 2 \times 3 = 6$ and $1 + 2 + 3 = 6$
6 is a *perfect number*.

perfect square a number that is the square of an integer
see *3·2 Square Roots*

Example: 25 is a *perfect square* since $25 = 5^2$.

perimeter the distance around the outside of a closed figure
see *Formulas page 58, 5·3 Evaluating Expressions and Formulas, 6·4 Perimeter, 8·4 Spreadsheets*

Example:

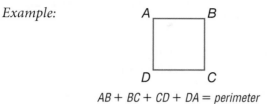

AB + BC + CD + DA = perimeter

permutation a possible arrangement of a group of objects; the number of possible arrangements of *n* objects is expressed by the term *n*! see *factorial, 4·5 Combinations and Permutations*

perpendicular two lines or planes that intersect to form a right angle see *6·5 Area*

Example:

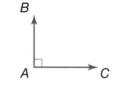

$\overline{AB}$ and $\overline{AC}$ are *perpendicular.*

pi the ratio of a circle's circumference to its diameter; *pi* is shown by the symbol π, and is approximately equal to 3.14 see *6·8 Circles, 8·1 Scientific Calculator*

picture graph a display of data that uses pictures or symbols to represent numbers see *4·2 Displaying Data*

place value the value given to a place a digit occupies in a numeral see *1·1 Place Value of Whole Numbers, 2·5 Naming and Ordering Decimals*

place-value system a number system in which values are given to the places digits occupy in the numeral; in the decimal system, the value of each place is 10 times the value of the place to its right

point one of four undefined terms in geometry used to define all other terms; a *point* has no size *see 5·7 Graphing on the Coordinate Plane, 6·1 Naming and Classifying Angles and Triangles*

polygon a simple, closed plane figure, having three or more line segments as sides *see 6·2 Polygons and Polyhedrons*

Examples:

polygons

polyhedron a solid geometrical figure that has four or more plane faces *see 6·2 Polygons and Polyhedrons*

Examples:

polyhedrons

population the universal set from which a sample of statistical data is selected *see 4·1 Collecting Data*

positive integers the set of all integers that are greater than zero {1, 2, 3, 4, 5, . . .} *see 1·5 Integer Operations*

positive numbers the set of all real numbers that are greater than zero {1, 1.36, $\sqrt{2}$, π}

power represented by the exponent n, to which a number is used as a factor n times *see 3·1 Powers and Exponents, 7·1 Systems of Measurement, 8·2 Scientific Calculator*

Example: 7 raised to the fourth *power.*
$$7^4 = 7 \times 7 \times 7 \times 7 = 2,401$$

predict to anticipate a trend by studying statistical data

prime factorization the expression of a composite number as a product of its prime factors *see 1·4 Factors and Multiples*

Examples: $504 = 2^3 \times 3^2 \times 7$
$30 = 2 \times 3 \times 5$

prime number a whole number greater than 1 whose only factors are 1 and itself *see 1·4 Factors and Multiples*

Examples: 2, 3, 5, 7, 11

prism a solid figure that has two parallel, congruent polygonal faces (called *bases*) *see 6·2 Polygons and Polyhedrons*

Examples:

prisms

probability the study of likelihood or chance that describes the possibility of an event occurring *see 4·4 Probability*

probability line a line used to order the probability of events from least likely to most likely *see 4·4 Probability*

product the result obtained by multiplying two numbers or variables *see 1·4 Factors and Multiples, 2·4 Multiplication and Division of Fractions, 5·1 Writing Expressions and Equations*

profit the gain from a business; what is left when the cost of goods and of carrying on the business is subtracted from the amount of money taken in

proportion a statement that two ratios are equal *see 2·8 Using and Finding Percents, 5·5 Ratio and Proportion*

pyramid a solid geometrical figure that has a polygonal base and triangular faces that meet at a common vertex *see 6·2 Polygons and Polyhedrons*

Examples:

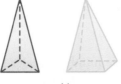

pyramids

Q

---

quadrant [1] one quarter of the circumference of a circle; [2] on a coordinate graph, one of the four regions created by the intersection of the *x*-axis and the *y*-axis *see 5·7 Graphing on the Coordinate Plane*

quadrilateral a polygon that has four sides *see 6·2 Polygons and Polyhedrons*

Examples:

quadrilaterals

qualitative graphs a graph with words that describes such things as a general trend of profits, income, and expenses over time; it has no specific numbers

quantitative graphs a graph that, in contrast to a qualitative graph, has specific numbers

quotient the result obtained from dividing one number or variable (the divisor) into another number or variable (the dividend) *see 1·5 Integer Operations, 5·1 Writing Expressions and Equations*

Example:

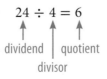

$$24 \div 4 = 6$$

dividend | quotient

divisor

radical the indicated root of a quantity *see 3·2 Square Roots*

 Examples: $\sqrt{3}, \sqrt[4]{14}, \sqrt[12]{23}$

radical sign the root symbol $\sqrt{}$

radius a line segment from the center of a circle to any point on its perimeter *see 6·8 Circles, 8·1 Four-Function Calculator*

random sample a population sample chosen so that each member has the same probability of being selected *see 4·1 Collecting Data*

range in statistics, the difference between the largest and smallest values in a sample *see 4·3 Statistics*

rank to order the data from a statistical sample on the basis of some criterion—for example, in ascending or descending numerical order *see 4·3 Statistics*

rate [1] fixed ratio between two things; [2] a comparison of two different kinds of units, for example, miles per hour or dollars per hour *see 5·5 Ratio and Proportion*

ratio a comparison of two numbers *see 2·7 Meaning of Percent, 5·5 Ratio and Proportion, 7·6 Size and Scale*

 Example: The *ratio* of consonants to vowels in the alphabet is 21:5.

rational numbers the set of numbers that can be written in the form $\frac{a}{b}$, where a and b are integers and b does not equal zero

 Examples: $1 = \frac{1}{1}, \frac{2}{9}, 3\frac{2}{7} = \frac{23}{7}, -0.333 = -\frac{1}{3}$

ray the part of a straight line that extends infinitely in one direction from a fixed point *see 6·1 Naming and Classifying Angles and Triangles, 8·3 Geometry Tools*

 Example:

 a *ray*

real numbers the set consisting of zero, all positive numbers, and all negative numbers; *real numbers* include all rational and irrational numbers

reciprocal one of a pair of numbers that have a product of 1
see 2·4 Multiplication and Division of Fractions, 8·1 Scientific Calculator
Examples: The *reciprocal* of 2 is $\frac{1}{2}$; of $\frac{3}{4}$ is $\frac{4}{3}$; of x is $\frac{1}{x}$.

rectangle a parallelogram with four right angles
see 6·2 Polygons and Polyhedrons
Example:

a *rectangle*

rectangular prism a prism that has rectangular bases and four rectangular faces *see 6·2 Polygons and Polyhedrons*

reflection a transformation that produces the mirror image of a figure *see 6·3 Symmetry and Transformations*
Example:

the *reflection* of a trapezoid

reflex angle any angle with a measure that is greater than 180° but less than 360° *see 6·1 Naming and Classifying Angles and Triangles*
Example:

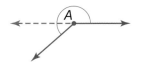

∠*A* is a *reflex angle.*

regular polygon a polygon in which all sides are equal and all angles are congruent *see 6·2 Polygons and Polyhedrons*

relationship a connection between two or more objects, numbers, or sets; a mathematical *relationship* can be expressed in words or with numbers and letters

repeating decimal a decimal in which a digit or a set of digits repeat infinitely *see 2·9 Fraction, Decimal, and Percent Relationships*

Example: 0.121212 . . . is a repeating decimal.

rhombus a parallelogram with all sides of equal length *see 6·2 Polygons and Polyhedrons*

Example:

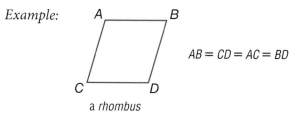

$AB = CD = AC = BD$

a *rhombus*

right angle an angle that measures 90° *see 6·1 Naming and Classifying Angles and Triangles*

Example:

∠A is a *right angle*.

right triangle a triangle with one right angle *see 6·1 Naming and Classifying Angles and Triangles*

Example:

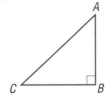

△*ABC* is a *right triangle.*

rise the vertical distance between two points

Roman numerals the numeral system consisting of the symbols I (1), V (5), X (10), L (50), C (100), D (500), and M (1,000); when a Roman symbol is preceded by a symbol of equal or greater value, the values of a symbol are added (XVI = 16); when a symbol is preceded by a symbol of lesser value, the values are subtracted (IV = 4)

root [1] the inverse of an exponent; [2] the radical sign $\sqrt{}$ indicates square root *see 3·2 Square Roots, 8·2 Scientific Calculator*

rotation a transformation in which a figure is turned a certain number of degrees around a fixed point or line *see 6·3 Symmetry and Transformations*

Example:

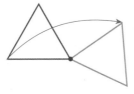

the *turning* of a triangle

round to approximate the value of a number to a given decimal place *see 1·1 Place Value of Whole Numbers*

> *Examples:* 2.56 rounded to the nearest tenth is 2.6;
> 2.54 rounded to the nearest tenth is 2.5;
> 365 rounded to the nearest hundred is 400.

row a horizontal list of numbers or terms *see 8·3 Spreadsheets*

rule a statement that describes a relationship between numbers or objects

run the horizontal distance between two points

· **S** ·

sample a finite subset of a population, used for statistical analysis *see 4·1 Collecting Data*

scale the ratio between the actual size of an object and a proportional representation *see 7·6 Size and Scale*

scale drawing a proportionally correct drawing of an object or area at actual, enlarged, or reduced size *see 7·6 Size and Scale*

scale factor the factor by which all the components of an object are multiplied in order to create a proportional enlargement or reduction *see 7·6 Size and Scale*

scalene triangle a triangle with no sides of equal length *see 6·1 Naming and Classifying Angles and Triangles*

> *Example:*

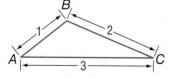

△*ABC* is a *scalene triangle.*

scale size the proportional size of an enlarged or reduced representation of an object or area *see 7·6 Size and Scale*

scatter plot (or scatter diagram) a display of data in which the points corresponding to two related factors are graphed and observed for correlation

Example:

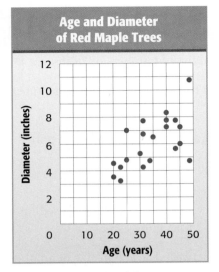

Age and Diameter of Red Maple Trees

scatter plot

segment two points and all the points on the line between them *see 6·1 Naming and Classifying Angles and Triangles, 6·8 Circles*

sequence *see Patterns page 62*

series *see Patterns page 62*

set a collection of distinct elements or items

side a line segment that forms an angle or joins the vertices of a polygon *see 6·1 Naming and Classifying Angles and Triangles, 6·4 Perimeter*

sighting measuring a length or angle of an inaccessible object by lining up a measuring tool with one's line of vision

signed number a number preceded by a positive or negative sign *see 1·5 Integer Operations*

significant digit the number of digits in a value that indicate its precision and accuracy

Example: 297,624 rounded to three significant digits is 298,000; 2.97624 rounded to three significant digits is 2.98.

similar figures figures that have the same shape but are not necessarily the same size *see 7·6 Size and Scale*

Example:

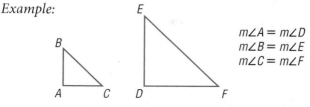

$m\angle A = m\angle D$
$m\angle B = m\angle E$
$m\angle C = m\angle F$

△*ABC* and △*DEF* are *similar figures.*

simple event an outcome or collection of outcomes *see 4·5 Probability*

simulation a mathematical experiment that approximates real-world processes

slide to move a shape to another position without rotating or reflecting it; also referred to as a translation *see 6·3 Symmetry and Transformations*

Example:

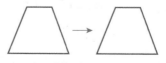

the *slide* of a trapezoid

slope [1] a way of describing the steepness of a line, ramp, hill, and so on; [2] the ratio of the rise to the run *see 6·8 Slope and Intercept*

slope angle the angle that a line forms with the *x*-axis or other horizontal

slope ratio the slope of a line as a ratio of the rise to the run

solid a three-dimensional figure *see 7·3 Area, Volume, and Capacity*

solution the answer to a mathematical problem; in algebra, a *solution* usually consists of a value or set of values for a variable *see 5·4 Equations*

speed the rate at which an object moves

speed-time graph a graph used to chart how the speed of an object changes over time

sphere a perfectly round geometric solid, consisting of a set of points equidistant from a center point

Example:

a *sphere*

spinner a device for determining outcomes in a probability experiment *see 4·4 Probability*

Example:

a *spinner*

spiral *see Patterns page 63*

spreadsheet a computer tool where information is arranged into cells within a grid and calculations are performed within the cells; when one cell is changed, all other cells that depend on it automatically change *see 8·3 Spreadsheets*

square [1] a rectangle with congruent sides *see 6·2 Polygons and Polyhedrons* [2] the product of two equal terms *see 3·1 Powers and Exponents, 7·1 Systems of Measurement, 8·2 Scientific Calculator*

Examples: [1]

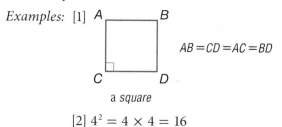

$$AB = CD = AC = BD$$

a *square*

[2] $4^2 = 4 \times 4 = 16$

square centimeter a unit used to measure the area of a surface; the area of a square measuring one centimeter on each side *see 7·3 Area, Volume, and Capacity*

square foot a unit used to measure the area of a surface; the area of a square measuring one foot on each side *see 7·3 Area, Volume, and Capacity*

square inch a unit used to measure the area of a surface; the area of a square measuring one inch on each side *see 7·3 Area, Volume, and Capacity*

square meter a unit used to measure the area of a surface; the area of a square measuring one meter on each side *see 7·3 Area, Volume, and Capacity*

square number *see Patterns page 63*

square pyramid a pyramid with a square base *see 6·2 Polygons and Polyhedrons*

square root a number that when multiplied by itself equals a given number *see 3·2 Square Roots, 8·1 Four-Function Calculator*

Example: 3 is the *square root* of 9.
$$\sqrt{9} = 3$$

square root symbol the mathematical symbol $\sqrt{}$; indicates that the square root of a given number is to be calculated *see 3·2 Square Roots*

standard measurement commonly used measurements, such as the meter used to measure length, the kilogram used to measure mass, and the second used to measure time *see Chapter 7 Measurement*

statistics the branch of mathematics that investigates the collection and analysis of data *see 4·3 Statistics*

steepness a way of describing the amount of incline (or slope) of a ramp, hill, line, and so on

stem the tens digit of an item of numerical data between 1 and 99 *see stem-and-leaf plot, 4·2 Displaying Data*

stem-and-leaf plot a method of displaying numerical data between 1 and 99 by separating each number into its tens digit (stem) and its unit digit (leaf) and then arranging the data in ascending order of the tens digits *see 4·2 Displaying Data*

Example:

Average Points per Game

Stem	Leaf
0	6
1	1 8 2 2 5
2	6 1
3	7
4	3
5	8

2 | 6 = 26 points

a *stem-and-leaf plot* for the data set
11, 26, 18, 12, 12, 15, 43, 37, 58, 6, and 21

straight angle an angle that measures 180°; a straight line *see 6·1 Naming and Classifying Angles and Triangles*

Subtraction Property of Equality the mathematical rule that states that if the same number is subtracted from each side of the equation, then the two sides remain equal *see 6·4 Solving Linear Equations*

Example: If $a = b$, then $a - c = b - c$.

sum the result of adding two numbers or quantities
see 5·1 Writing Expressions and Equations

Example: $6 + 4 = 10$

10 is the *sum* of the two addends, 6 and 4.

supplementary angles two angles that have measures whose sum is 180°

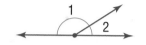

∠1 and ∠2 are *supplementary angles*.

surface area the sum of the areas of all the faces of a geometric solid, measured in square units *see 6·5 Surface Area*

Example:

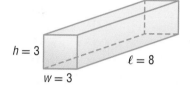

The *surface area* of this rectangular prism is
$2(3 \times 3) + 4(3 \times 8) = 114$ square units.

survey a method of collecting statistical data in which people are asked to answer questions *see 4·1 Collecting Data*

symmetry *see line of symmetry, 6·3 Symmetry and Transformations*

Example:

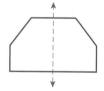

This hexagon has *symmetry* around the dotted line.

················· **T** ·················

table a collection of data arranged so that information can be easily seen *see 4·2 Collecting Data*

tally marks marks made for certain numbers of objects in keeping account *see 4·1 Collecting Data*

Example: ℍ‖ ‖‖ = 8

term product of numbers and variables *see 5·1 Writing Expressions and Equations*

Example: x, ax^2, $2x^4y^2$, and $-4ab$ are all *terms*.

terminating decimal a decimal with a finite number of digits *see 2·9 Fraction, Decimal, and Percent Relationships*

tessellation *see Patterns page 63*

tetrahedron a geometrical solid that has four triangular faces *see 6·2 Polygons and Polyhedrons*

Example:

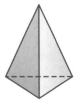

a *tetrahedron*

theoretical probability the ratio of the number of favorable outcomes to the total number of possible outcomes *see 4·4 Probability*

three-dimensional having three measurable qualities: length, height, and width

tiling completely covering a plane with geometric shapes *see tessellations*

time in mathematics, the element of duration, usually represented by the variable t *see 7·5 Time*

total distance the amount of space between a starting point and an endpoint, represented by d in the equation $d = s$ (speed) $\times t$ (time)

total distance graph a coordinate graph that shows cumulative distance traveled as a function of time

total time the duration of an event, represented by t in the equation $t = \dfrac{d \text{ (distance)}}{s \text{ (speed)}}$

transformation a mathematical process that changes the shape or position of a geometric figure *see 6·3 Symmetry and Transformations*

translation a transformation in which a geometric figure is slid to another position without rotation or reflection *see 6·3 Symmetry and Transformations*

trapezoid a quadrilateral with only one pair of parallel sides *see 6·2 Polygons and Polyhedrons*

Example:

a *trapezoid*

tree diagram a connected, branching graph used to diagram probabilities or factors *see 1·4 Factors and Multiples, 4·4 Probability*

Example:

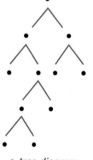

a *tree diagram*

trend a consistent change over time in the statistical data representing a particular population

triangle a polygon with three angles and three sides *see 6·1 Classifying Angles and Triangles*

triangular numbers *see Patterns page 63*

triangular prism a prism with two triangular bases and three rectangular sides *see 6·2 Polygons and Polyhedrons, 6·6 Surface Area*

turn to move a geometric figure by rotating it around a point *see 6·3 Symmetry and Transformations*

Example:

the *turning* of a triangle

two-dimensional having two measurable qualities: length and width

---------------------------------- **U** ----------------------------------

unequal probabilities different likelihoods of occurrence; two events have *unequal probabilities* if one is more likely to occur than the other

unfair where the probability of each outcome is not equal

union a set that is formed by combining the members of two or more sets, as represented by the symbol ∪; the *union* contains all members previously contained in both sets *see Venn diagram*

Example:

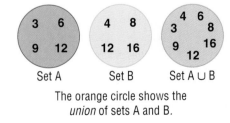

Set A   Set B   Set A ∪ B

The orange circle shows the *union* of sets A and B.

unit price the price of a single item or amount

unit rate the rate in lowest terms *see 5·5 Ratio and Proportion*

Example: 120 miles in two hours is equivalent to a *unit rate* of 60 miles per hour.

variable a letter or other symbol that represents a number or set of numbers in an expression or an equation *see 5·1 Writing Expressions and Equations*

Example: In the equation $x + 2 = 7$, the variable is x.

Venn diagram a pictorial means of representing the relationships between sets *see 1·4 Factors and Multiples*

Example:

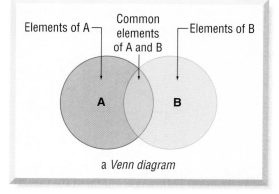

Elements of A — Common elements of A and B — Elements of B

A   B

a *Venn diagram*

vertex (pl. *vertices*) the common point of two rays of an angle, two sides of a polygon, or three or more faces of a polyhedron *see 6·1 Naming and Classifying Angles and Triangles, 8·3 Geometry Tools*

Examples:

vertex of an angle   *vertices* of a triangle   *vertices* of a cube

vertex of tessellation the point where three or more tessellating figures come together

Example:

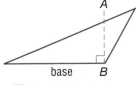

vertex of tessellation
(in the circle)

vertical a line that is perpendicular to a horizontal base line
see 5·7 Graphing on the Coordinate Plane, 8·4 Spreadsheets

Example:

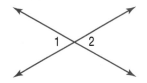

base $\quad$ B

$\overline{AB}$ is *vertical* to the base
of this triangle.

vertical angles opposite angles formed by the intersection of two lines *see 6·1 Naming and Classifying Angles and Triangles*

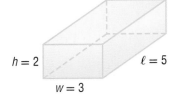

∠1 and ∠2 are *vertical angles.*

volume the space occupied by a solid, measured in cubic units
see Formulas page 58, 3·1 Powers and Exponents, 6·7 Volume, 7·3 Area, Volume, and Capacity

Example:

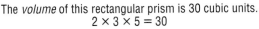

$h = 2$ $\qquad\qquad\qquad$ $\ell = 5$

$w = 3$

The *volume* of this rectangular prism is 30 cubic units.
$2 \times 3 \times 5 = 30$

W

whole numbers the set of all counting numbers plus zero {0, 1, 2, 3, 4, 5} *see 1·1 Place Value of Whole Numbers, 2·1 Fractions and Equivalent Fractions*

width a measure of the distance of an object from side to side *see 7·3 Area, Volume, and Capacity*

X

x-axis the horizontal reference line in the coordinate graph *see 5·7 Graphing on the Coordinate Plane*

x-intercept the point at which a line or curve crosses the x-axis

Y

y-axis the vertical reference line in the coordinate graph *see 5·7 Graphing on a Coordinate Plane*

y-intercept the point at which a line or curve crosses the y-axis

Z

zero-pair one positive cube and one negative cube used to model signed-number arithmetic

Formulas

Area *(see 6·5)*

circle	$A = \pi r^2$ (pi × square of the radius)
parallelogram	$A = bh$ (base × height)
rectangle	$A = \ell w$ (length × width)
square	$A = s^2$ (side squared)
trapezoid	$A = \frac{1}{2}h(b_1 + b_2)$
	($\frac{1}{2}$ × height × sum of the bases)
triangle	$A = \frac{1}{2}bh$
	($\frac{1}{2}$ × base × height)

Volume *(see 6·7)*

cone	$V = \frac{1}{3}\pi r^2 h$
	($\frac{1}{3}$ × pi × square of the radius × height)
cylinder	$V = \pi r^2 h$
	(pi × square of the radius × height)
prism	$V = Bh$ (area of the base × height)
pyramid	$V = \frac{1}{3}Bh$
	($\frac{1}{3}$ × area of the base × height)
rectangular prism	$V = \ell wh$ (length × width × height)
sphere	$V = \frac{4}{3}\pi r^3$
	($\frac{4}{3}$ × pi × cube of the radius)

Perimeter *(see 6·4)*

parallelogram	$P = 2a + 2b$
	(2 × side a + 2 × side b)
rectangle	$P = 2\ell + 2w$ (twice length + twice width)
square	$P = 4s$
	(4 × side)
triangle	$P = a + b + c$ (side a + side b + side c)

Formulas

Circumference *(see 6·8)*

circle $C = \pi d$ (pi × diameter)

or

$C = 2\pi r$

(2 × pi × radius)

Probability *(see 4·4)*

The *Experimental Probability* of an event is equal to the total number of times a favorable outcome occurred, divided by the total number of times the experiment was done.

$$\textit{Experimental Probability} = \frac{\textit{favorable outcomes that occurred}}{\textit{total number of experiments}}$$

The *Theoretical Probability* of an event is equal to the number of favorable outcomes, divided by the total number of possible outcomes.

$$\textit{Theoretical Probability} = \frac{\textit{favorable outcomes}}{\textit{possible outcome}}$$

Other

Distance $d = rt$ (rate × time)

Interest $I = prt$ (principle × rate × time)

PIE   Profit = Income − Expenses

Temperature $F = \frac{9}{5}C + 32$

$(\frac{9}{5} \times$ Temperature in °C $+ 32)$

$C = \frac{5}{9}(F - 32)$

$(\frac{5}{9} \times$ (Temperature in °F $- 32))$

Symbols

{ }	set	$\overline{AB}$	segment AB
∅	the empty set	$\overrightarrow{AB}$	ray AB
⊆	is a subset of	$\overleftrightarrow{AB}$	line AB
∪	union	$\triangle ABC$	triangle ABC
∩	intersection	$\angle ABC$	angle ABC
>	is greater than	$m\angle ABC$	measure of angle ABC
<	is less than		
≥	is greater than or equal to	$\overline{AB}$ or $m\overline{AB}$	length of segment AB
≤	is less than or equal to	$\overparen{AB}$	arc AB
=	is equal to	!	factorial
≠	is not equal to	$_nP_r$	permutations of n things taken r at a time
°	degree		
%	percent	$_nC_r$	combinations of n things taken r at a time
$f(n)$	function, f of n		
$a{:}b$	ratio of a to b, $\frac{a}{b}$	$\sqrt{}$	square root
$\lvert a \rvert$	absolute value of a	$\sqrt[3]{}$	cube root
$P(E)$	probability of an event E	′	foot
π	pi	″	inch
⊥	is perpendicular to	÷	divide
∥	is parallel to	/	divide
≅	is congruent to	*	multiply
∼	is similar to	×	multiply
≈	is approximately equal to	·	multiply
∠	angle	+	add
∟	right angle	−	subtract
△	triangle		

Patterns

arithmetic sequence a sequence of numbers or terms that have a common difference between any one term and the next in the sequence; in the following sequence, the common difference is seven, so $8 - 1 = 7$; $15 - 8 = 7$; $22 - 15 = 7$, and so forth

Example: 1, 8, 15, 22, 29, 36, 43, . . .

Fibonacci numbers a sequence in which each number is the sum of its two predecessors; can be expressed as $x_n = x_{n-2} + x_{n-1}$; the sequence begins: 1, 1, 2, 3, 5, 8, 13, 21, 34, 55, . . .

Example:

1,	1,	2,	3,	5,	8,	13,	21,	34,	55,	. . .
$1 + 1 = 2$										
	$1 + 2 = 3$									
		$2 + 3 = 5$								
			$3 + 5 = 8$							

geometric sequence a sequence of terms in which each term is a constant multiple, called the *common ratio,* of the one preceding it; for instance, in nature, the reproduction of many single-celled organisms is represented by a progression of cells splitting in two in a growth progression of 1, 2, 4, 8, 16, 32, . . ., which is a geometric sequence in which the common ratio is 2

harmonic sequence a progression $a_1, a_2, a_3, \ldots$ for which the reciprocals of the terms, $\frac{1}{a_1}, \frac{1}{a_2}, \frac{1}{a_3}, \ldots$ form an arithmetic sequence

Lucas numbers a sequence in which each number is the sum of its two predecessors; can be expressed as $x_n = x_{n-2} + x_{n-1}$; the sequence begins: 2, 1, 3, 4, 7, 11, 18, 29, 47, . . .

magic square a square array of different integers in which the sum of the rows, columns, and diagonals are the same

Example:

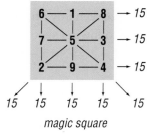

magic square

Pascal's triangle a triangular arrangement of numbers in which each number is the sum of the two numbers above it in the preceding row

Example:

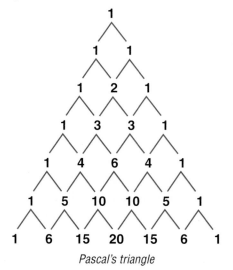

Pascal's triangle

sequence a set of elements, especially numbers, arranged in order according to some rule

series the sum of the terms of a sequence

spiral a plane curve traced by a point moving around a fixed point while continuously increasing or decreasing its distance from it

Example:

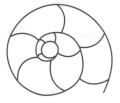

The shape of a chambered nautilus shell is a *spiral*.

square numbers a sequence of numbers that can be shown by dots arranged in the shape of a square; can be expressed as x^2; the sequence begins 1, 4, 9, 16, 25, 36, 49, . . .

Example:

square numbers

tessellation a tiling pattern made of repeating polygons that fills a plane completely, leaving no gaps

Example:

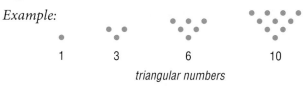

tessellations

triangular numbers a sequence of numbers that can be shown by dots arranged in the shape of a triangle; any number in the sequence can be expressed as $x_n = x_{n-1} + n$; the sequence begins 1, 3, 6, 10, 15, 21, . . .

Example:

1   3   6   10

triangular numbers

Hot Topics

HotTopic 1

Numbers and Computation

What do you know?

You can use the problems and the list of words that follow to see what you already know about this chapter. The answers to the problems are in **HotSolutions** at the back of the book, and the definitions of the words are in **HotWords** at the front of the book. You can find out more about a particular problem or word by referring to the topic number (*for example,* Lesson 1·2).

Problem Set

Give the value of the 3 in each number. (Lesson 1·1)

1. 237,514

2. 736,154,987

3. Write 24,378 using expanded form. (Lesson 1·1)

4. Write in order from greatest to least: 56,418; 566,418; 5,618; 496,418 (Lesson 1·1)

5. Round 52,564,764 to the nearest ten, thousand, and million. (Lesson 1·1)

Solve using mental math. (Lesson 1·2)

6. 258×0

7. $(5 \times 3) \times 1$

8. $4 \times (31 + 69)$

9. $25 \times 16 \times 4$

Use parentheses to make each expression true. (Lesson 1·3)

10. $4 + 6 \times 5 = 50$

11. $10 + 14 \div 3 + 3 = 4$

Is it a prime number? Write *yes* or *no*. (Lesson 1·4)

12. 99

13. 105

14. 106

15. 97

Write the prime factorization for each number. (Lesson 1·4)

16. 33

17. 105

18. 180

Find the GCF for each pair of numbers. (Lesson 1·4)

19. 15 and 30   **20.** 14 and 21   **21.** 18 and 120

Find the LCM for each pair of numbers. (Lesson 1·4)

22. 3 and 15   **23.** 12 and 8   **24.** 16 and 40

Write the opposite of each integer. (Lesson 1·5)

25. -6 **26.** 13   **27.** -15 **28.** 25

Add or subtract. (Lesson 1·5)

29. $9 + (-3)$ **30.** $4 - 5$ **31.** $-9 + (-9)$ **32.** $-8 - (-8)$

Compute. (Lesson 1·5)

33. $-4 \times (-7)$ **34.** $48 \div (-12)$ **35.** $-42 \div (-6)$

36. $(-4 \times 3) \times (-3)$ **37.** $3 \times [-6 + (-4)]$ **38.** $-5 [5 - (-7)]$

39. What can you say about the product of a negative integer and a positive integer? (Lesson 1·5)

40. What can you say about the sum of two positive integers? (Lesson 1·5)

HotWords

approximation (Lesson 1·1)
Associative Property (Lesson 1·2)
common factor (Lesson 1·4)
Commutative Property (Lesson 1·2)
composite number (Lesson 1·4)
Distributive Property (Lesson 1·2)
divisible (Lesson 1·4)
expanded form (Lesson 1·1)
factor (Lesson 1·4)
greatest common factor (Lesson 1·4)

least common multiple (Lesson 1·4)
multiple (Lesson 1·4)
negative integer (Lesson 1·5)
negative number (Lesson 1·5)
number system (Lesson 1·1)
operation (Lesson 1·3)
PEMDAS (Lesson 1·3)
place value (Lesson 1·1)
positive integer (Lesson 1·5)
prime factorization (Lesson 1·4)
prime number (Lesson 1·4)
round (Lesson 1·1)
Venn diagram (Lesson 1·4)

1·1 Place Value of Whole Numbers

Understanding Our Number System

Our **number system** is based on 10. The value of a digit is the product of that digit and its **place value**. For instance, in the number 5,700, the 5 has a value of five thousands and the 7 has a value of seven hundreds.

A *place-value chart* can help you read numbers. In the chart, each group of three digits is called a *period*. Commas separate the periods. The chart below shows the area of Asia, the largest continent. The area is about 17,300,000 square miles, which is nearly twice the size of North America.

Place-Value Chart

trillions period			billions period			millions period			thousands period			ones period		
hundred trillions	ten trillions	one trillions	hundred billions	ten billions	one billions	hundred millions	ten millions	one millions	hundred thousands	ten thousands	one thousands	hundreds	tens	ones
						1	7	3	0	0	0	0	0	0

To read a large number, think of the periods. At each comma, say the name of the period: 17,300,000 reads seventeen million, three hundred thousand.

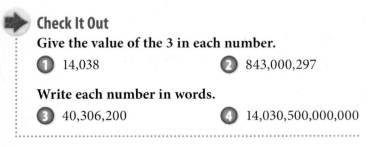

Check It Out

Give the value of the 3 in each number.

1 14,038

2 843,000,297

Write each number in words.

3 40,306,200

4 14,030,500,000,000

Using Expanded Form

To show the place values of the digits in a number, you can write the number using **expanded form**. You can write 50,203 using expanded form.

- Write the ten thousands. (5 × 10,000)
- Write the thousands. (0 × 1,000)
- Write the hundreds. (2 × 100)
- Write the tens. (0 × 10)
- Write the ones. (3 × 1)

So, 50,203 = (5 × 10,000) + (2 × 100) + (3 × 1).

Check It Out

Use expanded form to write each number.

5 83,046   **6** 300,285

Comparing and Ordering Numbers

When you compare numbers, there are exactly three possibilities: the first number is greater than the second (2 > 1); the second is greater than the first (3 < 4); or the two numbers are equal (6 = 6). When ordering several numbers, compare the numbers two at a time.

EXAMPLE Comparing Numbers

Compare 35,394 and 32,915.

35,394
32,915
    • Line up the digits, starting with the ones.

35,394
32,915
    • Beginning at the left, look at the digits in order. Find the first place where they differ. The digits in the thousands place differ.

5 > 2   • The number with the greater digit is greater.

So, 35,394 is greater than 32,915.

Check It Out

Write >, <, or =.

7 228,497 □ 238,006

8 52,004 □ 51,888

Write in order from least to greatest.

9 56,302; 52,617; 6,520; 526,000

Using Approximations

For many situations, using an **approximation** makes sense. For instance, it is reasonable to use a rounded number to express population. You might say that the population of a town is "about 60,000" rather than saying that it is "58,889."

Use this rule to **round** numbers. Look at the digit to the right of the place to which you are rounding. If the digit to the right is 5 or greater, round up. If it is less than 5, round down. Then replace all of the digits to the right of the rounded digit with zeros.

Round 123,456 to the nearest hundred.

Hundreds
↓
123,456
↑
$5 \geq 5$

So, 123,456 rounds to 123,500.

Check It Out

Round the given numbers.

10 Round 32,438 to the nearest hundred.

11 Round 558,925 to the nearest ten thousand.

12 Round 2,479,500 to the nearest million.

13 Round 369,635 to the nearest hundred thousand.

1·1 Exercises

Give the value of the 4 in each number.

1. 481,066

2. 628,014,257

Write each number in words.

3. 22,607,400

4. 3,040,680,000,000

Use expanded form to write each number.

5. 46,056

6. 4,800,325

Write >, <, or =.

7. 436,252 □ 438,352

8. 85,106 □ 58,995

Write in order from least to greatest.

9. 38,388; 83,725; 18,652; 380,735

Round 48,463,522 to each place indicated.

10. nearest ten

11. nearest thousand

12. nearest hundred thousand

13. nearest ten million

Solve.

14. In the first year, a video game had total sales of $226,520,000. During the second year, sales were $239,195,200. Did the game earn more money or less money in the second year? How do you know?

15. About 2,000,000 people visited the aquarium last year. If this number were rounded to the nearest million, what was the greatest number of visitors? What was the least number?

1·2 Properties

Commutative and Associative Properties

The operations of addition and multiplication share special properties because multiplication is repeated addition.

Both addition and multiplication are **commutative**. This means that the order doesn't change the sum or the product.
$$5 + 3 = 3 + 5 \text{ and } 5 \times 3 = 3 \times 5$$

If we let a and b be any whole numbers, then
$$a + b = b + a \text{ and } a \times b = b \times a.$$

Both addition and multiplication are **associative**. This means that grouping addends or factors will not change the sum or the product.
$$(5 + 7) + 9 = 5 + (7 + 9) \text{ and } (3 \times 2) \times 4 = 3 \times (2 \times 4).$$
$$(a + b) + c = a + (b + c) \text{ and } (a \times b) \times c = a \times (b \times c).$$

Subtraction and division do not share these properties. For example, neither is commutative.
$$(6 - 3) = 3, \text{ but } 3 - 6 = -3; \text{ therefore } 6 - 3 \neq 3 - 6.$$
$$6 \div 3 = 2, \text{ but } 3 \div 6 = 0.5; \text{ therefore } 6 \div 3 \neq 3 \div 6.$$

Similarly, we can show that neither subtraction nor division is associative.
$$(4 - 2) - 1 = 1; \text{ but } 4 - (2 - 1) = 3;$$
$$\text{therefore, } (4 - 2) - 1 \neq 4 - (2 - 1).$$
$$(4 \div 2) \div 2 = 1, \text{ but } 4 \div (2 \div 2) = 4;$$
$$\text{therefore, } (4 \div 2) \div 2 \neq 4 \div (2 \div 2).$$

Check It Out

Write *yes* or *no*.

1 $3 \times 7 = 7 \times 3$

2 $(8 \div 2) \div 2 = 8 \div (2 \div 2)$

3 $10 - 5 = 5 - 10$

4 $4 + (5 + 6) = (4 + 5) + 6$

Properties of One and Zero

When you add 0 to any number, the sum is that number. This is called the *Identity Property of Addition.*

   $32 + 0 = 32$

When you multiply any number by 1, the product is that number. This is called the *Identity Property of Multiplication.*

   $32 \times 1 = 32$

The product of any number and 0 is 0. This is called the *Zero Property of Multiplication.*

   $32 \times 0 = 0$

Check It Out

Solve.

5 $24{,}357 \times 1$ **6** $99 + 0$

7 $6 \times (5 \times 0)$ **8** $(3 \times 0.5) \times 1$

Distributive Property

The **Distributive Property** is important because it combines both addition and multiplication. This property states that multiplying a sum by a number is the same as multiplying each addend by that number and then adding the two products.

$$3(8 + 2) = (3 \times 8) + (3 \times 2)$$

If we let *a, b,* and *c* be any whole numbers, then

$$a \times (b + c) = (a \times b) + (a \times c).$$

Check It Out

Rewrite each expression using the Distributive Property.

9 $3 \times (3 + 2)$

10 $(5 \times 8) + (5 \times 7)$

11 $4 \times (4 + 4)$

Shortcuts for Adding and Multiplying

You can use the properties to help you perform some computations mentally.

$$77 + 56 + 23 = (77 + 23) + 56 = 100 + 56 = 156$$

Use the Commutative
and Associative Properties.

$$4 \times 9 \times 25 = (4 \times 25) \times 9 = 100 \times 9 = 900$$

$$8 \times 340 = (8 \times 300) + (8 \times 40) = 2{,}400 + 320 = 2{,}720$$

Use the Distributive Property.

Check It Out

Solve using mental math.

12 $25 \times 3 \times 8$

13 $3 \times (6 + 27 + 4)$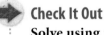

14 8×98

APPLICATION **Number Palindromes**

Do you notice anything interesting about this word, name, and question?

noon   Otto

Was it a rat I saw?

Each one is a *palindrome*—a word, name, or sentence that reads the same forward and backward. It is easy to make up number palindromes using three or more digits, such as 323 or 7227. But it is harder to make up a number sentence that is the same when you read its digits from either direction, such as $10989 \times 9 = 98901$. Try it and see!

1•2 Exercises

Write *yes* or *no*.

1. $7 \times 21 = 21 \times 7$

2. $3 \times 4 \times 7 = 3 \times 7 \times 4$

3. $3 \times 140 = (3 \times 100) \times (3 \times 40)$

4. $b \times (p + r) = bp + br$

5. $(2 \times 3 \times 5) = (2 \times 3) + (2 \times 5)$

6. $a \times (c + d + e) = ac + ad + ae$

7. $11 - 6 = 6 - 11$

8. $12 \div 3 = 3 \div 12$

Solve.

9. $22{,}350 \times 1$

10. $278 + 0$

11. $4 \times (0 \times 5)$

12. $0 \times 3 \times 15$

13. 0×1

14. $2.8 + 0$

15. 4.25×1

16. $(3 + 6 + 5) \times 1$

Rewrite each expression using the Distributive Property.

17. $5 \times (8 + 4)$

18. $(8 \times 12) + (8 \times 8)$

19. 4×350

Solve using mental math.

20. $5 \times (27 + 3)$

21. $6 \times (21 + 79)$

22. 7×220

23. 25×8

24. $2 + 63 + 98$

25. $150 + 50 + 450$

26. 130×6

27. $12 \times 50 \times 2$

28. Give an example to show that subtraction is not associative.

29. Give an example to show that division is not commutative.

30. Give an example to show the Zero (or Identity) Property of Addition.

1·3 Order of Operations

Understanding the Order of Operations

Solving a problem may involve using more than one **operation**. Your answer can depend on the order in which you do those operations.

For instance, consider the expression $2 + 3 \times 4$.

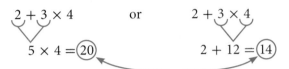

The order in which you perform operations makes a difference.

To make sure that there is just one answer to a series of computations, mathematicians have agreed on an order in which to do the operations.

EXAMPLE   Using the Order of Operations

Simplify $2 + 8 \times (9 - 5)$.

$2 + 8 \times (9 - 5)$
$\qquad = 2 + 8 \times 4$

- Simplify within the parentheses. Evaluate any powers. (See p. 158.)

$2 + 8 \times 4 = 2 + 32$

- Multiply or divide from left to right.

$2 + 32 = 34$

- Add or subtract from left to right.

So, $2 + 8 \times (9 - 5) = 34$.

➡️ **Check It Out**
Simplify.

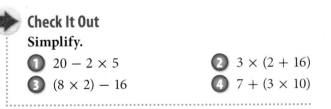

① $20 - 2 \times 5$

② $3 \times (2 + 16)$

③ $(8 \times 2) - 16$

④ $7 + (3 \times 10)$

1·3 Exercises

Is each expression true? Write *yes* or *no*.

1. $6 \times 3 + 4 = 22$
2. $3 + 6 \times 5 = 45$
3. $4 \times (6 + 4 \div 2) = 20$
4. $25 - (12 \times 1) = 13$
5. $(1 + 5) \times (1 + 5) = 36$
6. $(4 + 3 \times 2) + 6 = 20$
7. $35 - 5 \times 5 = 10$
8. $(9 \div 3) \times 9 = 27$

Simplify.

9. $24 - (3 \times 6)$
10. $3 \times (4 + 16)$
11. $2 \times 2 \times (8 - 5)$
12. $9 + (5 - 3)$
13. $(12 - 9) \times 5$
14. $10 + 9 \times 4$
15. $(4 + 5) \times 9$
16. $36 \div (12 + 6)$
17. $32 - (10 - 5)$
18. $24 + 6 \times (16 \div 2)$

Use parentheses to make the expression true.

19. $4 + 5 \times 6 = 54$
20. $4 \times 25 + 25 = 200$
21. $24 \div 6 + 2 = 3$
22. $10 + 20 \div 4 - 5 = 10$
23. $8 + 3 \times 3 = 17$
24. $16 - 10 \div 2 \times 4 = 44$

25. Use each number 2, 3, and 4 once to make an expression equal to 14.

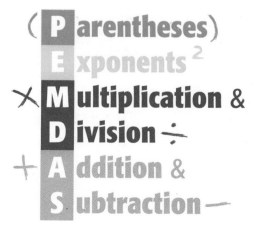

1·4 Factors and Multiples

Factors

Suppose that you want to arrange 15 small squares into a rectangular pattern. The only two options are shown below.

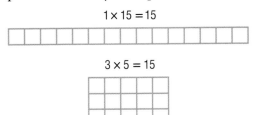

$1 \times 15 = 15$

$3 \times 5 = 15$

Two numbers multiplied together to produce 15 are considered **factors** of 15. So, the factors of 15 are 1, 3, 5, and 15.

To decide whether one number is a factor of another, divide. If there is a remainder of 0, the number is a factor.

EXAMPLE **Finding the Factors of a Number**

What are the factors of 20?

- Find all pairs of numbers that multiply to give the product.

    $1 \times 20 = 20$ $2 \times 10 = 20$ $4 \times 5 = 20$

- List the factors in order, starting with 1.

The factors of 20 are 1, 2, 4, 5, 10, and 20.

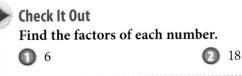

Check It Out

Find the factors of each number.

1 6   **2** 18

Common Factors

Factors that are the same for two or more numbers are called **common factors**.

EXAMPLE Finding Common Factors

What numbers are factors of both 8 and 20?

1, 2, 4, 8	• List the factors of the first number.
1, 2, 4, 5, 10, 20	• List the factors of the second number.
1, 2, 4	• Common factors are the numbers that are in both lists.

The common factors of 8 and 20 are 1, 2, and 4.

Check It Out

List the common factors of each set of numbers.

3 8 and 12   **4** 10, 15, and 20

Greatest Common Factor

The **greatest common factor** (GCF) of two whole numbers is the greatest number that is a factor of both the numbers.

EXAMPLE Finding the GCF

What is the GCF of 12 and 40?

• The factors of 12 are 1, 2, 3, 4, 6, 12.
• The factors of 40 are 1, 2, 4, 5, 8, 10, 20, 40.
• The common factors that are in both lists are 1, 2, 4.

The GCF of 12 and 40 is 4.

Check It Out

Find the GCF for each pair of numbers.

5 8 and 10   **6** 10 and 40

Venn Diagrams

A **Venn diagram** can be used to show how the elements of two or more collections of objects or numbers are related. Each collection, or *set*, is shown in a circle. Non-overlapping circles for sets A and B show that the sets have no elements in common.

When the circles in a Venn diagram overlap, as in sets C and D, the overlapping part contains the elements that are in both sets.

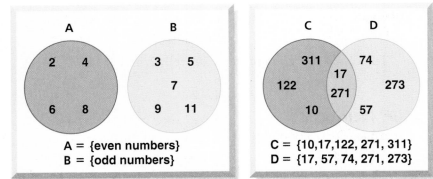

A = {even numbers}
B = {odd numbers}

C = {10, 17, 122, 271, 311}
D = {17, 57, 74, 271, 273}

You can use a Venn diagram to represent common factors of two or more numbers.

EXAMPLE Using a Venn Diagram to Show Common Factors

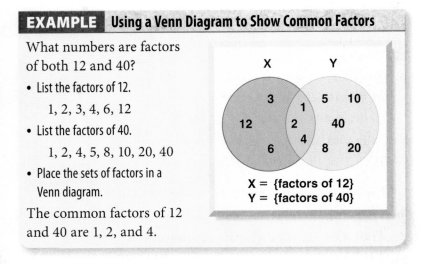

What numbers are factors of both 12 and 40?

• List the factors of 12.

    1, 2, 3, 4, 6, 12

• List the factors of 40.

    1, 2, 4, 5, 8, 10, 20, 40

• Place the sets of factors in a Venn diagram.

The common factors of 12 and 40 are 1, 2, and 4.

X = {factors of 12}
Y = {factors of 40}

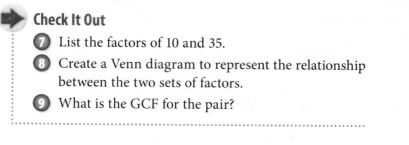

Check It Out

7 List the factors of 10 and 35.

8 Create a Venn diagram to represent the relationship between the two sets of factors.

9 What is the GCF for the pair?

Divisibility Rules

Sometimes you want to know whether a number is a factor of a much larger number. For instance, if you want to form teams of 3 from a group of 147 basketball players, you need to know whether 147 is *divisible* by 3. A number is **divisible** by another number if their quotient has no remainder.

You can quickly figure out whether 147 is divisible by 3 if you know the divisibility rule for 3. A number is divisible by 3 if the sum of the digits is divisible by 3. For example, 147 is divisible by 3 because $1 + 4 + 7 = 12$, and 12 is divisible by 3.

It can be helpful to know other divisibility rules. A number is divisible by:

2 if the ones digit is 0 or an even number.

3 if the sum of the digits is divisible by 3.

4 if the number formed by the last two digits is divisible by 4.

5 if the ones digit is 0 or 5.

6 if the number is divisible by 2 and 3.

8 if the number formed by the last three digits is divisible by 8.

9 if the sum of the digits is divisible by 9.

And . . .

Any number is divisible by **10** if the ones digit is 0.

Check It Out

10 Is 416 divisible by 4?   11 Is 129 divisible by 9?

12 Is 462 divisible by 6?   13 Is 1,260 divisible by 5?

Prime and Composite Numbers

A **prime number** is a whole number greater than 1 with exactly two factors, itself and 1. Here are the first 10 prime numbers:

2, 3, 5, 7, 11, 13, 17, 19, 23, 29

Twin primes are pairs of primes which have a difference of 2. (3, 5), (5, 7), and (11, 13) are examples of twin primes.

A number with more than two factors is called a **composite number**. When two composite numbers have no common factors (other than 1), they are said to be *relatively prime*. The numbers 8 and 25 are relatively prime. The factors of 8 are 1, 2, 4, and 8. The factors of 25 are 1, 5, and 25. Both 8 and 25 are composite numbers because both have more than two factors. However, they have no factor in common other than 1.

One way to find out whether a number is prime or composite is to use the sieve of Eratosthenes. Here is how it works.

- Use a chart of numbers listed in order. *Skip the number 1 because it is neither prime nor composite.*
- Circle the number 2, and then cross out all other multiples of 2.
- Next, circle the number 3, and then cross out all other multiples of 3.
- Then continue this procedure with 5, 7, 11, and with each succeeding number that has not been crossed out.
- The prime numbers are all the circled ones. The crossed-out numbers are the composite numbers.

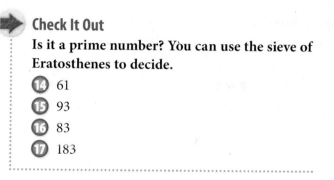

Check It Out

Is it a prime number? You can use the sieve of Eratosthenes to decide.

14 61

15 93

16 83

17 183

Prime Factorization

Every composite number can be expressed as a product of prime factors. Use a factor tree to find the prime factors.

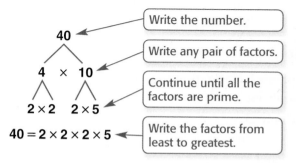

40 — Write the number.

4 × 10 — Write any pair of factors.

2 × 2   2 × 5 — Continue until all the factors are prime.

40 = 2 × 2 × 2 × 5 — Write the factors from least to greatest.

Although the order of the factors may be different because you can start with different pairs of factors, every factor tree for 40 has the same **prime factorization**. You can also write the prime factorization using exponents.

$$40 = 2 \times 2 \times 2 \times 5 = 2^3 \times 5$$

Check It Out

What is the prime factorization of each number?

18 30

19 80

20 120

21 110

Shortcut to Finding GCF

You can use prime factorization to find the greatest common factor.

> **EXAMPLE** Using Prime Factorization to Find the GCF
>
> Find the greatest common factor of 30 and 45.
>
> $30 = 2 \times 3 \times 5$
> $45 = 3 \times 3 \times 5$ • Find the prime factors of each number. Use a factor tree if it helps you.
>
> 3 and 5 • Find the prime factors common to both numbers.
>
> $3 \times 5 = 15$ • Find their product.
>
> The GCF of 30 and 45 is 3×5, or 15.

Check It Out

Use prime factorization to find the GCF of each pair of numbers.

22 6 and 15

23 10 and 30

24 12 and 30

25 24 and 36

Multiples and Least Common Multiples

The **multiples** of a number are the whole-number products when that number is a factor. In other words, you can find a multiple of a number by multiplying it by 1, 2, 3, and so on.

The **least common multiple** (LCM) of two numbers is the smallest nonzero number that is a multiple of both.

One way to find the LCM of a pair of numbers is to first list multiples of each and then identify the smallest one common to both. For instance, to find the LCM of 6 and 8:

- List the multiples of 6: 6, 12, 18, 24, 30, . . .
- List the multiples of 8: 8, 16, 24, 32, 40, . . .
- LCM = 24

Another way to find the LCM is to use prime factorization.

EXAMPLE **Using Prime Factorization to Find the LCM**

Find the least common multiple of 6 and 8.

$6 = 2 \times 3$
$8 = 2 \times 2 \times 2$

- Find the prime factorization of each number.

$2 \times 2 \times 2 \times 3 = 24$

- Multiply the prime factors of the lesser number by the prime factors of the greater number that are not factors of the least number.

The least common multiple of 6 and 8 is 24.

Check It Out

Use either method to find the LCM.

26 6 and 9
27 10 and 25
28 8 and 14
29 15 and 50

1·4 Exercises

Find the factors of each number.

1. 9

2. 24

3. 30

4. 48

Is it a prime number? Write *yes* or *no*.

5. 51

6. 79

7. 103

8. 219

Write the prime factorization for each number.

9. 55

10. 100

11. 140

12. 200

Find the GCF for each pair of numbers.

13. 8 and 24

14. 9 and 30

15. 18 and 25

16. 20 and 25

17. 16 and 30

18. 15 and 40

Find the LCM for each pair of numbers.

19. 6 and 7

20. 12 and 24

21. 16 and 24

22. 10 and 35

23. What is the divisibility rule for 6? Is 4,124 divisible by 8?

24. How do you use prime factorization to find the GCF of two numbers?

25. What is the least common multiple of 3, 4, and 5?

1·5 Integer Operations

Positive and Negative Integers

A glance through any newspaper shows that many quantities are expressed using **negative numbers**. For example, negative numbers show below-zero temperatures.

Whole numbers greater than zero are called **positive integers**. Whole numbers less than zero are called **negative integers**.

Here is the set of all integers:

{. . . , −5, −4, −3, −2, −1, 0, 1, 2, 3, 4, 5, . . .}

The integer 0 is neither positive nor negative. A number that has no sign is assumed to be positive.

 Check It Out

Write an integer to describe the situation.

1 3 below zero   **2** a gain of $250

Opposites of Integers

Integers can describe opposite ideas. Each integer has an opposite.

The opposite of a gain of 5 pounds is a loss of 5 pounds.
The opposite of +5 is −5.
The opposite of spending $3 is earning $3.
The opposite of −3 is +3.

 Check It Out

Write the opposite of each integer.

3 −12
4 +4
5 −8
6 0

Comparing and Ordering Integers

Remember that when you compare numbers, the first number is either equal to (=), greater than (>), or less than (<) the second number. You can use a number line to compare and order integers.

Comparing Integers

A lesser number appears to the left of a greater number on a number line.

$$\begin{array}{c} \leftarrow\!\!+\!\!+\!\!+\!\!+\!\!+\!\!+\!\!+\!\!+\!\!\bullet\!\!+\!\!+\!\!+\!\!+\!\!+\!\!+\!\!+\!\!\bullet\!\!+\!\!+\!\!+\!\!+\!\!+\!\!+\!\!+\!\!\rightarrow \\ {-10\,{-}9\,{-}8\,{-}7\,{-}6\,{-}5\,{-}4\,{-}3\,{-}2\,{-}1\ 0\ 1\ 2\ 3\ 4\ 5\ 6\ 7\ 8\ 9\ 10} \end{array}$$

2 is to the right of −4. Therefore, 2 > −4.
You can also write −4 < 2.

➡ **Check It Out**

Replace the ☐ with < or > to make a true sentence.

7 −2 ☐ −4

8 7 ☐ −1

9 −6 ☐ 0

10 3 ☐ −3

Ordering Integers

You can also use a number line to order several integers. Integers are ordered from least to greatest from left to right.

EXAMPLE Ordering Integers

Order 8, −1, −8, and 6 from least to greatest.

• Graph the numbers on a number line.

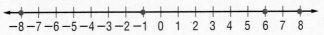

$$-8\,{-}7\,{-}6\,{-}5\,{-}4\,{-}3\,{-}2\,{-}1\ 0\ 1\ 2\ 3\ 4\ 5\ 6\ 7\ 8$$

• List the integers as they appear from left to right.

The order from least to greatest is −8, −1, 6, 8.

➡ **Check It Out**

11 Order 5, −3, −12, and 25 from least to greatest.

12 Order −1, 9, −8, and 6 from greatest to least.

Adding and Subtracting Integers

You can use a number line to model adding and subtracting integers.

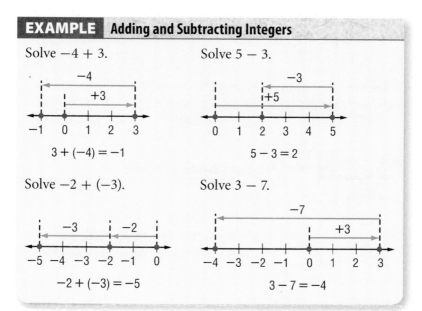

Adding or Subtracting Integers		
	Rules	**Examples**
Adding integers with the same sign	The sum of two positive integers is always positive. The sum of two negative integers is always negative.	$6 + 2 = 8$ $-6 + (-2) = -8$
Adding integers with different signs	The sum of a positive integer and a negative integer is sometimes positive, sometimes negative, and sometimes zero.	$6 + (-2) = 4$ $-6 + 2 = -4$ $-6 + 6 = 0$
Subtracting integers	To subtract an integer, add its opposite.	$6 - 1 = 6 + (-1)$ $-4 - 5 = -4 + (-5)$ $-8 - (-9) = -8 + 9$

➡ Check It Out

13 $5 - 7$ **14** $6 + (-6)$ **15** $-5 - (-7)$

16 $0 + (-3)$ **17** $-8 + 5$ **18** $-4 - 3$

Multiplying and Dividing Integers

Multiply and divide integers as you would whole numbers. Then use these rules for writing the sign of the answer.

The product of two integers with like signs is positive. The quotient is also positive.

$$2 \times 3 = 6 \qquad -4 \times (-3) = 12 \qquad -12 \div (-4) = 3$$

When the signs of the two integers are different, the product is negative. The quotient is also negative.

$$-6 \div 3 = -2 \qquad -3 \times 5 = -15 \qquad -4 \times 10 = -40$$

EXAMPLE **Multiplying and Dividing Integers**

Multiply $3 \times (-2)$.

$3 \times (-2)$ • Multiply the integers.

-6 • Remember that when the signs of the two integers are different, the product is negative.

So, $3 \times (-2) = -6$.

Divide $-8 \div 2$.

$-8 \div 2$ • Divide.

-4 • Remember that when the signs of the two integers are different, the quotient is negative.

So, $-8 \div 2 = -4$.

Check It Out

Find the product or quotient.

19 $-3 \times (-2)$

20 $12 \div (-4)$

21 $-15 \div (-3)$

22 -6×9

1·5 Exercises

Write the opposite of each integer.

1. -11 **2.** 5

3. -5 **4.** 2

Add or subtract.

5. $4 - 3$ **6.** $4 + (-6)$

7. $-5 - (-4)$ **8.** $0 + (-3)$

9. $-2 + 6$ **10.** $0 - 8$

11. $0 - (-6)$ **12.** $-3 - 8$

13. $7 + (-7)$ **14.** $-5 - (-8)$

15. $-2 - (-2)$ **16.** $-6 + (-9)$

Find the product or quotient.

17. $-2 \times (-6)$ **18.** $8 \div (-4)$

19. $-35 \div 5$ **20.** -5×7

21. $4 \times (-9)$ **22.** $-40 \div 8$

23. $-18 \div (-3)$ **24.** $6 \times (-7)$

Compute.

25. $[-6 \times (-2)] \times 3$ **26.** $4 \times [2 \times (-4)]$

27. $[-3 \times (-3)] \times -3$ **28.** $-4 \times [3 + (-4)]$

29. $[-7 \times (-3)] \times 4$ **30.** $-2 \times [6 - (-2)]$

31. What can you say about the sum of two negative integers?

32. The temperature at noon was 10°F. For the next 3 hours it dropped at a rate of 3 degrees an hour. First express this change as an integer. Then give the temperature at 3:00 P.M.

33. What can you say about the product of a positive integer and a negative integer?

Numbers and Computation

What have you learned?

You can use the problems and the list of words that follow to see what you learned in this chapter. You can find out more about a particular problem or word by referring to the topic number (*for example,* Lesson 1·2).

Problem Set

Give the value of the 8 in each number. (Lesson 1·1)

1. 287,617

2. 758,122,907

3. Write 36,514 using expanded form. (Lesson 1·1)

4. Write in order from greatest to least: 243,254; 283,254; 83,254; and 93,254 (Lesson 1·1)

5. Round 46,434,482 to the nearest ten, thousand, and million. (Lesson 1·1)

Solve. (Lesson 1·2)

6. 736×0 **7.** $(5 \times 4) \times 1$ **8.** $5,945 + 0$ **9.** 0×0

Solve using mental math. (Lesson 1·2)

10. $8 \times (34 + 66)$ **11.** $50 \times 15 \times 2$

Use parentheses to make each expression true. (Lesson 1·3)

12. $5 + 7 \times 2 = 24$ **13.** $32 + 12 \div 4 + 5 = 40$

Is it a prime number? Write *yes* or *no*. (Lesson 1·4)

14. 51   **15.** 102   **16.** 173   **17.** 401

Write the prime factorization for each number. (Lesson 1·4)

18. 35   **19.** 130   **20.** 190

Find the GCF for each pair of numbers. (Lesson 1·4)

21. 16 and 36   **22.** 12 and 45   **23.** 20 and 160

Find the LCM for each pair of numbers. (Lesson 1·4)

24. 5 and 10   **25.** 12 and 8   **26.** 18 and 20

27. What is the divisibility rule for 10? Is 2,530 a multiple of 10? (Lesson 1·4)

Write the opposite of each integer. (Lesson 1·5)

28. -4 **29.** 14   **30.** -17 **31.** -5

Add or subtract. (Lesson 1·5)

32. $10 + (-9)$ **33.** $3 - 8$ **34.** $-4 + (-4)$
35. $2 - (-2)$ **36.** $-12 - (-12)$ **37.** $-6 + 12$

Compute. (Lesson 1·5)

38. $-9 \times (-6)$ **39.** $36 \div (-12)$
40. $-54 \div (-9)$ **41.** $(-4 \times 2) \times (-5)$
42. $3 \times [-6 + (-6)]$ **43.** $-2 + [4 - (-9)]$

44. What can you say about the quotient of a positive integer and a negative integer? (Lesson 1·5)

HotWords

Write definitions for the following words.

approximation (Lesson 1·1)
Associative Property
(Lesson 1·2)
common factor (Lesson 1·4)
Commutative Property
(Lesson 1·2)
composite number (Lesson 1·4)
Distributive Property
(Lesson 1·2)
divisible (Lesson 1·4)
expanded form (Lesson 1·1)
factor (Lesson 1·4)
greatest common factor
(Lesson 1·4)

least common multiple
(Lesson 1·4)
multiple (Lesson 1·4)
negative integer (Lesson 1·5)
negative number (Lesson 1·5)
number system (Lesson 1·1)
operation (Lesson 1·3)
PEMDAS (Lesson 1·3)
place value (Lesson 1·1)
positive integer (Lesson 1·5)
prime factorization (Lesson 1·4)
prime number (Lesson 1·4)
round (Lesson 1·1)
Venn diagram (Lesson 1·4)

HotTopic 2

Fractions, Decimals, and Percents

What do you know?

You can use the problems and the list of words that follow to see what you already know about this chapter. The answers to the problems are in **HotSolutions** at the back of the book, and the definitions of the words are in **HotWords** at the front of the book. You can find out more about a particular problem or word by referring to the topic number (*for example,* Lesson 2·1).

Problem Set

1. For a family vacation, Chenelle bought 3 polo shirts for $6.75 each and 3 hats for $8.50 each. How much money did she spend? (Lesson 2·6)

2. Kelly got 4 out of 50 problems wrong on her social studies test. What percent did she get correct? (Lesson 2·8)

3. Which fraction is not equivalent to $\frac{2}{3}$? (Lesson 2·1)

   A. $\frac{4}{6}$ B. $\frac{40}{60}$ C. $\frac{12}{21}$ D. $\frac{18}{27}$

Add or subtract. Write your answers in simplest form. (Lesson 2·3)

4. $3\frac{2}{3} + \frac{1}{2}$ 5. $2\frac{1}{4} - \frac{4}{5}$ 6. $6 - 2\frac{1}{6}$ 7. $2\frac{1}{7} + 2\frac{4}{9}$

Solve. Write your answers in simplest form. (Lesson 2·4)

8. $\frac{5}{6} \times \frac{3}{8}$ 9. $\frac{3}{5} \div 4\frac{1}{3}$

10. Give the place value of 6 in 23.064. (Lesson 2·5)

11. Write in expanded form: 4.603. (Lesson 2·5)

12. Write as a decimal: two hundred forty-seven thousandths. (Lesson 2·5)

13. Write the following numbers in order from least to greatest: 1.655; 1.605; 16.5; 1.065. (Lesson 2·5)

Find each answer as indicated. (Lesson 2·6)

14. 5.466 + 12.45   **15.** 13.9 − 0.677   **16.** 4.3 × 23.67

Use a calculator to answer Exercises 17 and 18. Round to the nearest tenth. (Lesson 2·8)

17. What percent of 56 is 14?   **18.** Find 16% of 33.

Write each decimal as a percent. (Lesson 2·9)

19. 0.68   **20.** 0.5

Write each fraction as a percent. (Lesson 2·9)

21. $\frac{6}{100}$ **22.** $\frac{56}{100}$

Write each percent as a decimal. (Lesson 2·9)

23. 34%   **24.** 125%

HotWords

benchmark (Lesson 2·7)

common denominator (Lesson 2·2)

compatible numbers (Lesson 2·3)

cross product (Lesson 2·1)

denominator (Lesson 2·1)

equivalent (Lesson 2·1)

equivalent fractions (Lesson 2·1)

estimate (Lesson 2·3)

factor (Lesson 2·4)

fraction (Lesson 2·1)

greatest common factor (Lesson 2·1)

improper fraction (Lesson 2·1)

inverse operations (Lesson 2·4)

least common multiple (Lesson 2·2)

mixed number (Lesson 2·1)

numerator (Lesson 2·1)

percent (Lesson 2·7)

place value (Lesson 2·5)

product (Lesson 2·4)

proportion (Lesson 2·8)

ratio (Lesson 2·7)

reciprocal (Lesson 2·4)

repeating decimal (Lesson 2·9)

terminating decimal (Lesson 2·9)

whole number (Lesson 2·1)

2·1 Fractions and Equivalent Fractions

Naming Fractions

A **fraction** can be used to name a part of a whole. The flag of Sierra Leone is divided into three equal parts: green, white, and blue. Each part, or color, of the flag represents $\frac{1}{3}$ of the whole flag. $\frac{3}{3}$, or 1, represents the whole flag.

A fraction can also name part of a set. There are four balls in the set shown. Each ball is $\frac{1}{4}$ of the set. $\frac{4}{4}$, or 1, equals the whole set. Three of the balls are baseballs. The baseballs represent $\frac{3}{4}$ of the set. One of the four balls is a football. The football represents $\frac{1}{4}$ of the set.

You name fractions by their **numerators** and **denominators**.

EXAMPLE Naming Fractions

Write a fraction for the number of shaded rectangles.

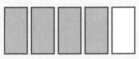

- The denominator of the fraction tells the number of equal parts of the whole set.

  There are 5 rectangles altogether.
- The numerator of the fraction tells the number of parts under consideration.

  There are 4 shaded rectangles.
- Write the fraction:

$$\frac{\text{parts under consideration}}{\text{parts that make the whole set}} = \frac{\text{numerator}}{\text{denominator}}$$

The fraction for the number of shaded rectangles is $\frac{4}{5}$.

Check It Out

Write the fraction for each picture.

1 ___ of the circle is shaded.

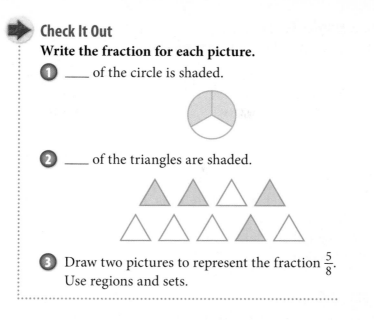

2 ___ of the triangles are shaded.

3 Draw two pictures to represent the fraction $\frac{5}{8}$. Use regions and sets.

Methods for Finding Equivalent Fractions

Equivalent fractions are fractions that describe the same amount of a region or set. You can use fraction pieces to show equivalent fractions.

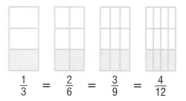

$$\frac{1}{3} = \frac{2}{6} = \frac{3}{9} = \frac{4}{12}$$

Each of the fraction pieces represents fractions equal to $\frac{1}{3}$. This makes them equivalent fractions.

Fraction Names for One

An infinite number of fractions are equal to 1.

Names for One				Not Names for One			
$\frac{2}{2}$	$\frac{365}{365}$	$\frac{1}{1}$	$\frac{5}{5}$	$\frac{1}{0}$	$\frac{3}{1}$	$\frac{1}{365}$	$\frac{11}{12}$

Any number multiplied by one is still equal to the original number, so knowing different names for 1 can help you find equivalent fractions.

To find a fraction that is **equivalent** to another fraction, you can multiply the original fraction by a form of 1. You can also divide the numerator and denominator by the same number to get an equivalent fraction.

EXAMPLE **Methods for Finding Equivalent Fractions**

Find a fraction equal to $\frac{9}{12}$.

• Multiply the fraction by a form of 1, or divide the numerator and denominator by the same number.

<table>
<tr><th colspan="1">Multiply</th><th>OR</th><th>Divide</th></tr>
</table>

$$\frac{9}{12} \times \frac{2}{2} = \frac{18}{24} \qquad\qquad \frac{9 \div 3}{12 \div 3} = \frac{3}{4}$$

$$\frac{9}{12} = \frac{18}{24} \qquad\qquad\qquad \frac{9}{12} = \frac{3}{4}$$

So, $\frac{9}{12} = \frac{18}{24} = \frac{3}{4}$.

 Check It Out

Write two equivalent fractions.

4 $\frac{1}{3}$

5 $\frac{6}{12}$

6 $\frac{3}{5}$

7 Write three fraction names for 1.

Deciding Whether Two Fractions Are Equivalent

Two fractions are equivalent if you can show that each fraction is just a different name for the same amount.

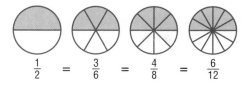

$$\frac{1}{2} = \frac{3}{6} = \frac{4}{8} = \frac{6}{12}$$

Another method you can use to identify equivalent fractions is to find the **cross products** of the fractions.

EXAMPLE Deciding Whether Two Fractions Are Equivalent

Determine whether $\frac{2}{3}$ is equivalent to $\frac{10}{15}$.

$\frac{2}{3} \overset{?}{\times} \frac{10}{15}$ • Cross multiply the fractions.

$2 \times 15 \overset{?}{=} 10 \times 3$ • Compare the cross products.

$30 = 30$ • If the cross products are the same, then the fractions are equivalent.

So, $\frac{2}{3} = \frac{10}{15}$.

Check It Out

Use the cross products method to determine whether each pair of fractions is equivalent.

8 $\frac{3}{4}, \frac{27}{36}$

9 $\frac{5}{6}, \frac{25}{30}$

10 $\frac{15}{32}, \frac{45}{90}$

Writing Fractions in Simplest Form

When the numerator and the denominator of a fraction have no common factor other than 1, the fraction is in *simplest form*. You can use fraction pieces to show fractions in equivalent forms and then identify the simplest.

$\frac{4}{8}$ and $\frac{1}{2}$ are equivalent fractions.

The common factors of 4 and 8 are 1, 2, and 4. Because the numerator and denominator of $\frac{4}{8}$ have common factors other than 1, the fraction is not in simplest form.

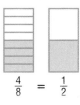

$$\frac{4}{8} = \frac{1}{2}$$

The numerator and denominator of $\frac{1}{2}$ have no common factor other than 1.

Also, the fewest number of fraction pieces to show the equivalent of $\frac{4}{8}$ is $\frac{1}{2}$. Therefore, the fraction $\frac{4}{8}$ is equal to $\frac{1}{2}$ in simplest form.

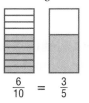

$$\frac{6}{10} = \frac{3}{5}$$

$\frac{6}{10}$ and $\frac{3}{5}$ are equivalent fractions. The numerator and denominator of $\frac{6}{10}$ have more common factors than 1.

The numerator and denominator of $\frac{3}{5}$ have no common factor other than 1. Therefore, $\frac{3}{5}$ is the fraction written in simplest form.

To express fractions in simplest form, you can divide the numerator and denominator by their **greatest common factor** (GCF).

EXAMPLE **Finding Simplest Form of Fractions**

Express $\frac{12}{18}$ in simplest form.

The factors of 12 are:
    1, 2, 3, 4, 6, 12

• List the factors of the numerator.

The factors of 18 are:
    1, 2, 3, 6, 9, 18

• List the factors of the denominator.

The greatest common factor of 12 and 18 is 6.

• Find the greatest common factor (GCF).

$\frac{12 \div 6}{18 \div 6} = \frac{2}{3}$

• Divide the numerator and the denominator of the fraction by the GCF.

$\frac{2}{3}$

• Write the fraction in simplest form.

So, $\frac{12}{18}$ expressed in simplest form is $\frac{2}{3}$.

Check It Out

Express each fraction in simplest form.

⑪ $\frac{4}{20}$

⑫ $\frac{9}{27}$

⑬ $\frac{18}{20}$

APPLICATION Musical Fractions

In music, notes are written on a series of lines called a *staff*. The shape of a note shows its time value—how long the note lasts when the music is played.

A whole note has the longest time value. A half note is held half as long as a whole note. Other notes are held for other fractions of time, compared with the whole note. Each flag makes the value of the note half of what it was before the flag was added.

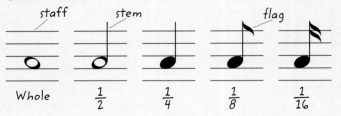

A series of short notes may be connected by a beam instead of writing each one with a flag.

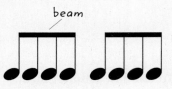

How long is each of these notes?

Write a series of notes to show that 1 whole note is equal to 2 half notes, 4 quarter notes, and 16 sixteenth notes.

See **HotSolutions** for answers.

Writing Improper Fractions and Mixed Numbers

You can write fractions for amounts greater than 1. A fraction with a numerator greater than or equal to the denominator is called an **improper fraction**.

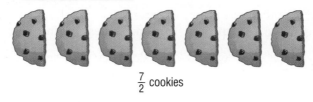

$\frac{7}{2}$ cookies

$\frac{7}{2}$ is an improper fraction.

A **whole number** and a fraction make up a **mixed number**.

$3\frac{1}{2}$ cookies

$3\frac{1}{2}$ is a mixed number.

You can write any mixed number as an improper fraction and any improper fraction as a mixed number. You can use division to change an improper fraction to a mixed number.

EXAMPLE **Changing an Improper Fraction to a Mixed Number**

Change $\frac{8}{3}$ to a mixed number.

- Divide the numerator by the denominator.

Divisor $\longrightarrow$ $3\overline{)8}$ 2 $\longleftarrow$ Quotient

$\underline{6}$

2 $\longleftarrow$ Remainder

- Write the mixed number.

Quotient $\longrightarrow$ $2\frac{2}{3}$ $\longleftarrow$ Remainder
$\longleftarrow$ Divisor

You can use multiplication to change a mixed number to an improper fraction. Start by renaming the whole-number part. Rename it as an improper fraction with the same denominator as the fraction part, and then add the two parts.

EXAMPLE **Changing a Mixed Number to an Improper Fraction**

Change $2\frac{1}{2}$ to an improper fraction. Express in simplest form.

$2 \times \frac{2}{2} = \frac{4}{2}$
- Multiply the whole-number part by a version of one that has the same denominator as the fraction part.

$2\frac{1}{2} = \frac{4}{2} + \frac{1}{2} = \frac{5}{2}$
- Add the two parts (p. 110).

So, $2\frac{1}{2} = \frac{5}{2}$.

Check It Out

Write a mixed number for each improper fraction.

⑭ $\frac{24}{5}$

⑮ $\frac{13}{9}$

⑯ $\frac{33}{12}$

⑰ $\frac{29}{6}$

Write an improper fraction for each mixed number.

⑱ $1\frac{7}{10}$

⑲ $5\frac{1}{8}$

⑳ $6\frac{3}{5}$

㉑ $7\frac{3}{7}$

2·1 Exercises

Write the fraction for each picture.

1. ____ of the fruits are lemons.

2. ____ of the circle is red.

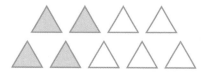

3. ____ of the triangles are green.

4. ____ of the balls are basketballs.

5. ____ of the balls are baseballs.

Write the fraction.

6. four ninths

7. twelve thirteenths

8. fifteen thirds

9. two halves

Write one fraction equivalent to the given fraction.

10. $\frac{2}{5}$

11. $\frac{1}{9}$

12. $\frac{9}{36}$

13. $\frac{60}{70}$

Express each fraction in simplest form.

14. $\frac{10}{24}$

15. $\frac{16}{18}$

16. $\frac{36}{40}$

Write each improper fraction as a mixed number.

17. $\frac{24}{7}$

18. $\frac{32}{5}$

19. $\frac{12}{7}$

Write each mixed number as an improper fraction.

20. $2\frac{3}{4}$

21. $11\frac{8}{9}$

22. $1\frac{5}{6}$

23. $3\frac{1}{3}$

24. $4\frac{1}{4}$

25. $8\frac{3}{7}$

2•2 Comparing and Ordering Fractions

Comparing Fractions

You can use fraction pieces to compare fractions.

$$\frac{1}{2} > \frac{2}{5}$$

$$\frac{1}{4} < \frac{5}{6}$$

You can also compare fractions if you find *equivalent fractions* (p. 97) and compare numerators.

EXAMPLE **Comparing Fractions**

Compare the fractions $\frac{4}{5}$ and $\frac{5}{7}$.

• Look at the denominators.

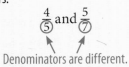

Denominators are different.

• If the denominators are different, write equivalent fractions with a **common denominator**.

35 is the least common multiple of 5 and 7.
Use 35 for the common denominator.

$$\frac{4}{5} \times \frac{7}{7} = \frac{28}{35} \qquad \frac{5}{7} \times \frac{5}{5} = \frac{25}{35}$$

• Compare the numerators.

28 > 25

• The fractions compare as the numerators compare.

$$\frac{28}{35} > \frac{25}{35}, \text{ so } \frac{4}{5} > \frac{5}{7}.$$

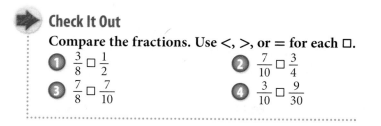

Check It Out

Compare the fractions. Use $<$, $>$, or $=$ for each □.

1 $\frac{3}{8} \square \frac{1}{2}$

2 $\frac{7}{10} \square \frac{3}{4}$

3 $\frac{7}{8} \square \frac{7}{10}$

4 $\frac{3}{10} \square \frac{9}{30}$

Comparing Mixed Numbers

To compare *mixed numbers* (p. 103), first compare the whole numbers. Then compare the fractions, if necessary.

EXAMPLE Comparing Mixed Numbers

Compare $1\frac{2}{5}$ and $1\frac{4}{7}$.

- Check that the fractions are not improper.

  $\frac{2}{5}$ and $\frac{4}{7}$ are not improper.

- Compare the whole-number parts. If they are different, the one that is greater is the greater mixed number. If they are equal, go on.

  $1 = 1$

- Compare the fraction parts by renaming them with a *common denominator*.

  35 is the least common multiple of 5 and 7.

  Use 35 for the common denominator.

  $\frac{2}{5} \times \frac{7}{7} = \frac{14}{35}$ $\frac{4}{7} \times \frac{5}{5} = \frac{20}{35}$

- Compare the fractions.

  $\frac{14}{35} < \frac{20}{35}$, so $1\frac{2}{5} < 1\frac{4}{7}$.

Check It Out

Compare each mixed number. Use $<$, $>$, or $=$ for each □.

5 $1\frac{3}{4} \square 1\frac{2}{5}$

6 $2\frac{2}{9} \square 2\frac{1}{17}$

7 $5\frac{16}{19} \square 5\frac{4}{7}$

Ordering Fractions

To compare and order fractions, you can find equivalent fractions and then compare the numerators of the fractions.

EXAMPLE Ordering Fractions with Unlike Denominators

Order the fractions $\frac{2}{5}$, $\frac{3}{4}$, and $\frac{3}{10}$ from least to greatest.

- Find the **least common multiple** (LCM) (p. 85) of $\frac{2}{5}$, $\frac{3}{4}$, and $\frac{3}{10}$.

  Multiples of 4: 4, 8, 12, 16, 20, 24, . . .
  Multiples of 5: 5, 10, 15, 20, 25, . . .
  Multiples of 10: 10, 20, 30, 40, . . .
  The LCM of 4, 5, and 10 is 20.

- Write equivalent fractions with the LCM as the common denominator. The LCM of the denominators of two fractions is called the *least common denominator* (LCD).

$$\frac{2}{5} = \frac{2}{5} \times \frac{4}{4} = \frac{8}{20}$$

$$\frac{3}{4} = \frac{3}{4} \times \frac{5}{5} = \frac{15}{20}$$

$$\frac{3}{10} = \frac{3}{10} \times \frac{2}{2} = \frac{6}{20}$$

- The fractions compare as the numerators compare.

$$\frac{6}{20} < \frac{8}{20} < \frac{15}{20}, \text{ so } \frac{3}{10} < \frac{2}{5} < \frac{3}{4}.$$

Check It Out

Order the fractions from least to greatest.

8 $\frac{2}{4}, \frac{4}{5}, \frac{5}{8}$

9 $\frac{3}{4}, \frac{2}{3}, \frac{7}{12}$

10 $\frac{5}{6}, \frac{2}{3}, \frac{5}{8}$

Find the LCD of each pair of fractions.

11 $\frac{3}{4}, \frac{5}{7}$

12 $\frac{5}{7}, \frac{3}{18}$

2·2 Exercises

Find the LCD for each pair of fractions.

1. $\frac{2}{3}, \frac{3}{5}$

2. $\frac{1}{4}, \frac{5}{12}$

Compare each fraction. Use <, >, or =.

3. $\frac{1}{2}, \frac{3}{7}$

4. $\frac{10}{12}, \frac{5}{6}$

5. $\frac{4}{5}, \frac{3}{4}$

6. $\frac{5}{8}, \frac{2}{3}$

7. $\frac{1}{5}, \frac{20}{100}$

8. $\frac{4}{3}, \frac{3}{4}$

Compare each mixed number. Use <, >, or = for each □.

9. $3\frac{3}{8} \ \square \ 3\frac{4}{7}$

10. $1\frac{2}{3} \ \square \ 1\frac{3}{5}$

11. $2\frac{3}{4} \ \square \ 2\frac{5}{6}$

12. $5\frac{4}{5} \ \square \ 5\frac{5}{8}$

13. $2\frac{1}{3} \ \square \ 2\frac{3}{9}$

14. $2\frac{3}{4} \ \square \ 1\frac{4}{5}$

Order the fractions and mixed numbers from least to greatest.

15. $\frac{4}{7}, \frac{1}{3}, \frac{9}{14}$

16. $\frac{2}{3}, \frac{5}{9}, \frac{4}{7}$

17. $\frac{3}{8}, \frac{1}{2}, \frac{5}{32}, \frac{3}{4}$

18. $\frac{2}{3}, \frac{5}{6}, \frac{5}{24}, \frac{7}{12}$

19. $2\frac{1}{3}, \frac{6}{3}, \frac{3}{4}, \frac{13}{4}$

20. $\frac{4}{5}, \frac{7}{10}, \frac{15}{5}, \frac{16}{10}$

Use the information about recess soccer goals to answer Exercise 21.

Recess Soccer Goals	
An-An	$\frac{2}{5}$
Derrick	$\frac{4}{7}$
Roberto	$\frac{5}{8}$
Gwen	$\frac{8}{10}$
numerator = goals made denominator = goals attempted	

21. Who was more accurate, Derrick or An-An?

22. The Wildcats won $\frac{3}{4}$ of their games. The Hawks won $\frac{5}{6}$ of theirs. The Bluejays won $\frac{7}{8}$ of theirs. Which team won the greatest fraction of their games? the least?

2·3 Addition and Subtraction of Fractions

Adding and Subtracting Fractions with Like Denominators

When you add or subtract fractions that have the same, or like, *denominators,* you add or subtract only the *numerators.* The denominator stays the same.

$$\frac{1}{3} + \frac{2}{3} = \frac{3}{3} = 1$$

EXAMPLE **Adding and Subtracting Fractions with Like Denominators**

Add $\frac{1}{8} + \frac{5}{8}$.

- Add or subtract the numerators.

  $\frac{1}{8} + \frac{5}{8}$ $1 + 5 = 6$

- Write the result over the denominator.

  $\frac{1}{8} + \frac{5}{8} = \frac{6}{8}$

- Simplify, if possible.

  $\frac{6}{8} = \frac{3}{4}$

So, $\frac{1}{8} + \frac{5}{8} = \frac{3}{4}$ and

Subtract $\frac{8}{10} - \frac{3}{10}$.

$\frac{8}{10} - \frac{3}{10}$ $8 - 3 = 5$

$\frac{8}{10} - \frac{3}{10} = \frac{5}{10}$

$\frac{5}{10} = \frac{1}{2}$

$\frac{8}{10} - \frac{3}{10} = \frac{1}{2}$.

Check It Out

Add or subtract. Express your answers in simplest form.

❶ $\frac{5}{6} + \frac{7}{6}$ ❷ $\frac{4}{25} + \frac{2}{25}$ ❸ $\frac{11}{23} - \frac{6}{23}$ ❹ $\frac{13}{16} - \frac{7}{16}$

Adding and Subtracting Fractions with Unlike Denominators

To add or subtract fractions with unlike denominators, you rename the fractions so that they have the same denominator.

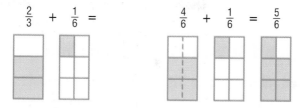

$$\frac{2}{3} + \frac{1}{6} = \qquad \frac{4}{6} + \frac{1}{6} = \frac{5}{6}$$

Change $\frac{2}{3}$ to sixths so that the fractions have the same denominator. Then add.

To add or subtract fractions with unlike denominators, you need to change them to equivalent fractions with common, or like, denominators before you find the sum or difference.

EXAMPLE **Adding Fractions with Unlike Denominators**

Add $\frac{1}{4} + \frac{3}{8}$.

- Find the *least common multiple* (LCM) (p. 85) of 4 and 8.

    Multiples of 8: 8, 16, 24, 32, . . .
    Multiples of 4: 4, 8, 12, 16, . . .
    The LCM of 4 and 8 is 8.

- Write equivalent fractions with the LCM as the common denominator.
    $$\frac{1}{4} \times \frac{2}{2} = \frac{2}{8} \text{ and } \frac{3}{8} = \frac{3}{8}$$
- Add the fractions. Express the fraction in simplest form.
    $$\frac{2}{8} + \frac{3}{8} = \frac{5}{8}$$
So, $\frac{1}{4} + \frac{3}{8} = \frac{5}{8}$.

Check It Out

Add or subtract. Express your answers in simplest form.

5 $\frac{3}{4} + \frac{1}{2}$

6 $\frac{5}{6} - \frac{2}{3}$

7 $\frac{1}{5} + \frac{1}{2}$

8 $\frac{2}{3} - \frac{1}{12}$

Adding and Subtracting Mixed Numbers

Adding and subtracting mixed numbers is similar to adding and subtracting fractions. Sometimes you have to rename your number to subtract. Sometimes you will have an improper fraction to simplify.

Adding Mixed Numbers with Common Denominators

To add *mixed numbers* with common denominators, you just need to write the sum of the numerators over the common denominator. Then add the whole numbers.

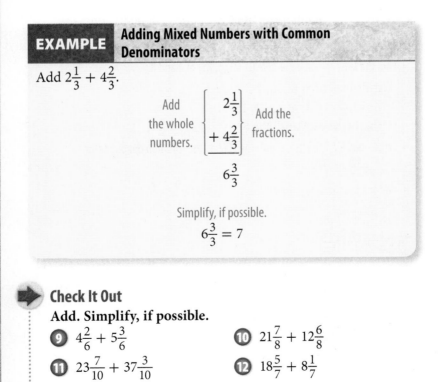

EXAMPLE Adding Mixed Numbers with Common Denominators

Add $2\frac{1}{3} + 4\frac{2}{3}$.

Add the whole numbers.
$$2\frac{1}{3}$$
$$+ 4\frac{2}{3}$$
Add the fractions.

$$6\frac{3}{3}$$

Simplify, if possible.
$$6\frac{3}{3} = 7$$

Check It Out

Add. Simplify, if possible.

9 $4\frac{2}{6} + 5\frac{3}{6}$

10 $21\frac{7}{8} + 12\frac{6}{8}$

11 $23\frac{7}{10} + 37\frac{3}{10}$

12 $18\frac{5}{7} + 8\frac{1}{7}$

Adding Mixed Numbers with Unlike Denominators

You can use fraction pieces to model the addition of mixed numbers with unlike denominators.

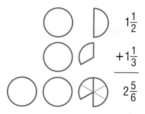

$1\frac{1}{2}$

$+1\frac{1}{3}$

$2\frac{5}{6}$

To add mixed numbers with unlike denominators, you need to write equivalent fractions with a common denominator.

EXAMPLE **Adding Mixed Numbers with Unlike Denominators**

Add $2\frac{2}{5} + 3\frac{1}{10}$.

- Write equivalent fractions with a common denominator.

$$2\frac{2}{5} = 2\frac{4}{10} \text{ and } 3\frac{1}{10} = 3\frac{1}{10}$$

- Add.

Add the whole numbers. $\left\{ \begin{array}{l} 2\frac{4}{10} \\ + 3\frac{1}{10} \end{array} \right.$ Add the fractions.

$$5\frac{5}{10}$$

Simplify, if possible.

$$5\frac{5}{10} = 5\frac{1}{2}$$

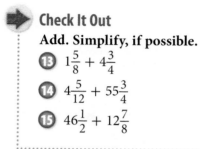

Check It Out

Add. Simplify, if possible.

13 $1\frac{5}{8} + 4\frac{3}{4}$

14 $4\frac{5}{12} + 55\frac{3}{4}$

15 $46\frac{1}{2} + 12\frac{7}{8}$

Subtracting Mixed Numbers with Common or Unlike Denominators

You can model the subtraction of mixed numbers with unlike denominators.

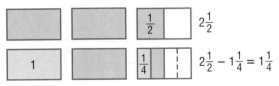

To subtract mixed numbers, you need to have or make common denominators.

EXAMPLE Subtracting Mixed Numbers

Subtract $12\frac{2}{3} - 5\frac{3}{4}$.

• If the denominators are unlike, write equivalent fractions with a common denominator.

$$12\frac{2}{3} \longrightarrow 12\frac{8}{12} \longrightarrow 11\frac{20}{12}$$
$$-5\frac{3}{4} \longrightarrow -5\frac{9}{12} \longrightarrow -5\frac{9}{12}$$
$$\overline{\phantom{-5\frac{9}{12}}} \qquad 6\frac{11}{12}$$

So, $12\frac{2}{3} - 5\frac{3}{4} = 6\frac{11}{12}$.

 Check It Out

Subtract. Express your answers in simplest form.

16 $6\frac{7}{8} - \frac{1}{2}$

17 $32\frac{1}{2} - 16\frac{5}{15}$

18 $30\frac{4}{5} - 12\frac{5}{6}$

19 $26\frac{2}{5} - 17\frac{7}{10}$

Estimating Fraction Sums and Differences

To **estimate** fraction sums and differences, you can use the estimation techniques of rounding or substituting **compatible numbers**. Compatible numbers are close to the real numbers in the problem but easier to add or subtract mentally.

EXAMPLE Estimating Fraction Sums and Differences

Estimate the sum of $7\frac{3}{8} + 8\frac{1}{9} + 5\frac{6}{7}$.

- **Substitute Compatible Numbers.**

  Change each number to a whole number or a mixed number containing $\frac{1}{2}$.

  $$7\frac{3}{8} + 8\frac{1}{9} + 5\frac{6}{7}$$
  $$\downarrow \quad\quad \downarrow \quad\quad \downarrow$$
  $$7\frac{1}{2} + 8 \quad + 6 = 21\frac{1}{2}$$

- **Round the Fraction Parts.**

  Round down if the fraction part is less than $\frac{1}{2}$. Round up if the fraction part is greater than or equal to $\frac{1}{2}$.

  $$7\frac{3}{8} + 8\frac{1}{9} + 5\frac{6}{7}$$
  $$\downarrow \quad\quad \downarrow \quad\quad \downarrow$$
  $$7 \quad + 8 \quad + 6 = 21$$

➡️ **Check It Out**

Estimate each sum or difference. Use both the compatible numbers method and the rounding method for each problem.

20 $6\frac{1}{4} - 2\frac{5}{6}$

21 $12\frac{1}{8} - 4\frac{3}{4}$

22 $2 + 1\frac{1}{2} + 5\frac{3}{8}$

23 $3\frac{1}{8} + \frac{3}{4} + 4\frac{1}{6}$

2·3 Exercises

Add or subtract. Express in simplest form.

1. $\dfrac{3}{16} + \dfrac{5}{16}$

2. $\dfrac{5}{25} + \dfrac{6}{25}$

3. $\dfrac{3}{8} + \dfrac{7}{8}$

4. $\dfrac{15}{29} - \dfrac{14}{29}$

5. $\dfrac{25}{60} - \dfrac{15}{60}$

Add or subtract. Express in simplest form.

6. $\dfrac{3}{5} + \dfrac{6}{9}$

7. $\dfrac{4}{12} + \dfrac{2}{15}$

8. $\dfrac{5}{18} + \dfrac{1}{6}$

9. $\dfrac{7}{9} - \dfrac{2}{3}$

10. $\dfrac{5}{9} - \dfrac{1}{3}$

Estimate each sum or difference.

11. $7\dfrac{9}{10} + 8\dfrac{3}{4}$

12. $4\dfrac{5}{6} - 3\dfrac{3}{4}$

13. $2\dfrac{9}{10} + 8\dfrac{3}{5} + 1\dfrac{1}{2}$

14. $13\dfrac{4}{5} - 6\dfrac{1}{9}$

15. $5\dfrac{1}{3} + 2\dfrac{7}{8} + 6\dfrac{1}{4}$

Add or subtract. Simplify, if possible.

16. $7\dfrac{4}{10} - 3\dfrac{1}{10}$

17. $3\dfrac{3}{8} - 1\dfrac{2}{8}$

18. $13\dfrac{4}{5} + 12\dfrac{2}{5}$

19. $24\dfrac{6}{11} + 11\dfrac{5}{11}$

20. $22\dfrac{2}{7} + 11\dfrac{4}{7}$

Add. Simplify, if possible.

21. $3\dfrac{1}{2} + 5\dfrac{1}{4}$

22. $17\dfrac{1}{3} + 23\dfrac{1}{6}$

23. $26\dfrac{3}{4} + 5\dfrac{1}{2}$

24. $21\dfrac{7}{10} + 16\dfrac{3}{5}$

Subtract. Simplify, if possible.

25. $6\dfrac{1}{5} - 1\dfrac{9}{10}$

26. $19\dfrac{1}{4} - 1\dfrac{1}{2}$

27. $48\dfrac{1}{3} - 19\dfrac{11}{12}$

28. $55\dfrac{3}{8} - 26\dfrac{2}{7}$

29. Maria is painting her bedroom. She has $4\dfrac{2}{3}$ gallons of paint. She needs $\dfrac{3}{4}$ gallon for the trim and $3\dfrac{1}{2}$ gallons for the walls. Does she have enough paint to paint her bedroom?

30. Maria has a piece of molding $22\dfrac{5}{8}$ feet long. She used $8\dfrac{2}{3}$ feet for one wall in the bedroom. Does she have enough molding for the other wall, which is $12\dfrac{1}{2}$ feet long?

2·4 Multiplication and Division of Fractions

Multiplying Fractions

You know that 2×2 means "2 groups of 2." Multiplying fractions involves the same concept: $2 \times \frac{1}{2}$ means "2 groups of $\frac{1}{2}$." You may find it helpful to think of *times* as *of* in word form.

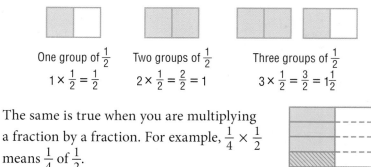

One group of $\frac{1}{2}$ Two groups of $\frac{1}{2}$ Three groups of $\frac{1}{2}$

$1 \times \frac{1}{2} = \frac{1}{2}$ $2 \times \frac{1}{2} = \frac{2}{2} = 1$ $3 \times \frac{1}{2} = \frac{3}{2} = 1\frac{1}{2}$

The same is true when you are multiplying a fraction by a fraction. For example, $\frac{1}{4} \times \frac{1}{2}$ means $\frac{1}{4}$ of $\frac{1}{2}$.

$\frac{1}{4} \times \frac{1}{2} = \frac{1}{8}$

When you are not using models to multiply fractions, you multiply the numerators and then the denominators. It is not necessary to find a common denominator.

$$\frac{1}{3} \times \frac{1}{4} = \frac{1}{12}$$

EXAMPLE Multiplying Fractions

Multiply $\frac{4}{5}$ and $\frac{5}{6}$.

$\frac{4}{5} \times \frac{5}{6}$

- Write mixed numbers, if any, as improper fractions (p. 101).

$\frac{4}{5} \times \frac{5}{6} = \frac{20}{30}$

- Multiply the numerators.
- Multiply the denominators.

$\frac{20 \div 10}{30 \div 10} = \frac{2}{3}$

- Write the product in simplest form, if necessary.

So, $\frac{4}{5} \times \frac{5}{6} = \frac{2}{3}$.

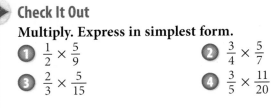

Check It Out

Multiply. Express in simplest form.

1 $\dfrac{1}{2} \times \dfrac{5}{9}$ **2** $\dfrac{3}{4} \times \dfrac{5}{7}$

3 $\dfrac{2}{3} \times \dfrac{5}{15}$ **4** $\dfrac{3}{5} \times \dfrac{11}{20}$

Shortcut for Multiplying Fractions

You can use a shortcut when you multiply fractions. Instead of multiplying across and then writing the product in simplest form, you can simplify before multiplying.

EXAMPLE   **Simplifying Before Multiplying**

Multiply $\dfrac{2}{5}$ and $\dfrac{10}{14}$.

$\dfrac{2}{5} \times \dfrac{10}{14}$ • Write mixed numbers, if any, as improper fractions.

$= \dfrac{2}{5} \times \dfrac{2 \times 5}{2 \times 7}$ • Show the factors of the numerators and denominators. Identify whether the numerators and denominators have any common factors.

$= \dfrac{\overset{1}{2} \times 2 \times 5}{5 \times \underset{1}{2} \times 7}$ • Divide both the numerator and the denominator by the common factors—first divide both by 2.

$= \dfrac{2 \times 2 \times \overset{1}{5}}{\underset{1}{5} \times 2 \times 7}$ • Divide both the numerator and denominator by 5.

$= \dfrac{1 \times 2 \times 1}{1 \times 1 \times 7}$ • Multiply.

$= \dfrac{2}{7}$ • Write the product in simplest form, if necessary.

Check It Out

Find common factors and simplify, and then multiply.

5 $\dfrac{4}{9} \times \dfrac{3}{8}$ **6** $\dfrac{3}{5} \times \dfrac{15}{21}$

7 $\dfrac{12}{15} \times \dfrac{4}{9}$ **8** $\dfrac{7}{8} \times \dfrac{16}{21}$

Finding the Reciprocal of a Number

The **reciprocal** of a number is the *inverse* of that number. Inverse operations are operations that *undo* each other. For example, multiplication and division are inverse operations. To find the reciprocal of a number, you switch the numerator and the denominator.

Number	Reciprocal
$\frac{4}{5}$	$\frac{5}{4}$
$3 = \frac{3}{1}$	$\frac{1}{3}$
$6\frac{1}{2} = \frac{13}{2}$	$\frac{2}{13}$

When you multiply a number by its reciprocal, the product is 1.
$$\frac{2}{5} \times \frac{5}{2} = \frac{10}{10} = 1$$

The number 0 does not have a reciprocal.

Check It Out

Find the reciprocal of each number.

9 $\frac{3}{8}$ **10** 5   **11** $4\frac{1}{2}$

Multiplying Mixed Numbers

You can use what you know about multiplying fractions to help you multiply mixed numbers. To multiply mixed numbers, you rewrite them as improper fractions.

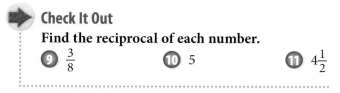

| EXAMPLE | Multiplying Mixed Numbers |

Multiply $2\frac{1}{3} \times 1\frac{1}{4}$.

$2\frac{1}{3} \times 1\frac{1}{4} = \frac{7}{3} \times \frac{5}{4}$

$\frac{7}{3} \times \frac{5}{4} = \frac{35}{12}$

$\frac{35}{12} = 2\frac{11}{12}$

- Write the mixed numbers as improper fractions.
- Cancel factors, if possible, and then multiply the fractions.
- Change to a mixed number and reduce to simplest form, if necessary.

Check It Out

Multiply. Express in simplest form.

⑫ $2\frac{2}{5} \times 3\frac{2}{6}$ ⑬ $4\frac{5}{9} \times 2\frac{1}{16}$ ⑭ $15\frac{2}{3} \times 4\frac{5}{8}$

Dividing Fractions

When you divide a fraction by another fraction, such as $\frac{1}{3} \div \frac{1}{6}$, you are actually finding out how many $\frac{1}{6}$ are in $\frac{1}{3}$.

$$\frac{1}{3} \div \frac{1}{6} = 2$$

Two $\frac{1}{6}$ are in $\frac{1}{3}$.

Dividing by a number is the same as multiplying by its reciprocal. So, to divide fractions, replace the divisor with its reciprocal and then multiply to get the answer.

$$\frac{1}{3} \div \frac{1}{6} = \frac{1}{3} \times \frac{6}{1} = 2$$

EXAMPLE Dividing Fractions

Divide $\frac{3}{4} \div \frac{9}{10}$.

$\frac{3}{4} \div \frac{9}{10} = \frac{3}{4} \times \frac{10}{9}$

- Replace the divisor with its reciprocal, and multiply.

$\frac{\overset{1}{3}}{2 \times \underset{1}{2}} \times \frac{\overset{1}{2 \times 5}}{\underset{1}{3 \times 3}} = \frac{1}{2} \times \frac{5}{3}$

- Simplify by dividing common factors.

$\frac{1}{2} \times \frac{5}{3} = \frac{5}{6}$

- Multiply the fractions.

So, $\frac{3}{4} \div \frac{9}{10} = \frac{5}{6}$.

Check It Out

Divide. Express in simplest form.

⑮ $\frac{3}{4} \div \frac{3}{5}$ ⑯ $\frac{5}{7} \div \frac{1}{2}$ ⑰ $\frac{7}{9} \div \frac{1}{8}$

Dividing Mixed Numbers

When you divide $3\frac{3}{4}$ by $1\frac{1}{4}$, you are actually finding out how many sets of $1\frac{1}{4}$ are in $3\frac{3}{4}$.

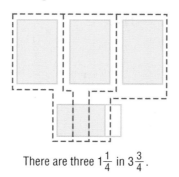

There are three $1\frac{1}{4}$ in $3\frac{3}{4}$.

Dividing mixed numbers is similar to dividing fractions, but first you need to change the *mixed numbers* to *improper fractions* (p. 104).

$$3\frac{3}{4} \div 1\frac{1}{4} = \frac{15}{4} \div \frac{5}{4} = \frac{15}{4} \times \frac{4}{5} = \frac{60}{20} = 3$$

EXAMPLE Dividing Mixed Numbers

Solve $2\frac{1}{2} \div 1\frac{1}{3}$.

$2\frac{1}{2} \div 1\frac{1}{3} = \frac{5}{2} \div \frac{4}{3}$

- Write the mixed numbers as improper fractions.

The reciprocal of $\frac{4}{3}$ is $\frac{3}{4}$.

- Replace the divisor with its reciprocal.

$\frac{5}{2} \times \frac{3}{4} = \frac{15}{8} = 1\frac{7}{8}$

- Multiply the fractions. Write in simplest form, if necessary.

So, $2\frac{1}{2} \div 1\frac{1}{3} = 1\frac{7}{8}$.

Check It Out

Divide. Express fractions in simplest form.

18 $1\frac{1}{8} \div \frac{3}{4}$ **19** $\frac{16}{2} \div 1\frac{1}{2}$ **20** $4\frac{3}{4} \div 6\frac{1}{3}$

Multiply. Write in simplest form.

1. $\frac{2}{5} \times \frac{7}{9}$

2. $\frac{3}{8} \times \frac{3}{5}$

3. $\frac{6}{7} \times \frac{7}{8}$

4. $\frac{3}{4} \times \frac{5}{11}$

5. $\frac{4}{7} \times \frac{21}{24}$

Find the reciprocal.

6. $\frac{5}{7}$

7. $5\frac{1}{2}$

8. 4

9. $6\frac{2}{3}$

10. $7\frac{5}{6}$

11. 27

Multiply. Write in simplest form.

12. $5\frac{1}{8} \times 12\frac{2}{7}$

13. $3\frac{3}{4} \times 16\frac{4}{5}$

14. $11\frac{1}{2} \times 4\frac{1}{6}$

15. $10\frac{2}{9} \times 2\frac{13}{16}$

16. $6\frac{5}{12} \times 4\frac{4}{9}$

17. $8\frac{4}{5} \times 5\frac{5}{8}$

Divide. Write in simplest form.

18. $\frac{1}{5} \div \frac{2}{3}$

19. $\frac{4}{9} \div \frac{11}{15}$

20. $\frac{3}{8} \div \frac{12}{21}$

21. $\frac{13}{19} \div \frac{26}{27}$

22. $\frac{21}{26} \div \frac{12}{13}$

Divide. Write in simplest form.

23. $5\frac{5}{6} \div 7\frac{7}{9}$

24. $3\frac{3}{5} \div 2\frac{2}{17}$

25. $12\frac{2}{7} \div 2\frac{13}{15}$

26. $7\frac{1}{3} \div 6\frac{1}{9}$

27. $4\frac{4}{5} \div 3\frac{4}{5}$

28. $3\frac{1}{3} \div 1\frac{2}{3}$

29. Each dinner at the Shady Tree Truck Stop is served with $\frac{1}{2}$ cup of corn. If there are about 4 cups of corn in a pound, about how many dinners could be served with 16 pounds of corn?

30. Jamie is making chocolate cupcakes. The recipe calls for $\frac{2}{3}$ cup of cocoa. She wants to make $\frac{1}{2}$ of the recipe. How much cocoa will she need?

2·5 Naming and Ordering Decimals

Decimal Place Value: Tenths and Hundredths

You can use what you know about **place value** of whole numbers to read and write decimals.

0.01   0.1   1

Place-Value Chart

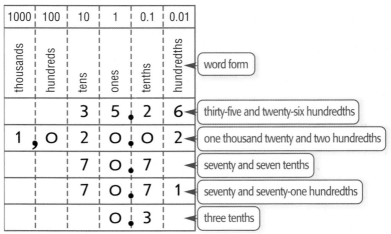

1000	100	10	1	0.1	0.01	
thousands	hundreds	tens	ones	tenths	hundredths	word form
		3	5 .	2	6	thirty-five and twenty-six hundredths
1 ,	0	2	0 .	0	2	one thousand twenty and two hundredths
		7	0 .	7		seventy and seven tenths
		7	0 .	7	1	seventy and seventy-one hundredths
			0 .	3		three tenths

You read a decimal by reading the whole number to the left of the decimal point as usual. You say "and" for the decimal point. Then find the place of the last decimal digit, and use it to name the decimal part.

You can use a place-value chart to help you read and write decimal numbers.

You can write a decimal in *word form* in *standard form,* or in *expanded form.*

Standard form is the most common way to write a decimal. You write the whole number, place the decimal point, and then write the last digit of the decimal number in the place that names it.

standard form | expanded form
70.71 | $(7 \times 10) + (0 \times 1) + (7 \times 0.1) + (1 \times 0.01)$

Expanded form is the sum of the products of each digit and its place value.

Decimal Place Value: Thousandths

One thousandth is the number 1 divided by 1,000. The base-ten blocks show that 100 thousandths is equal to 1 tenth and that 10 thousandths is equal to 1 hundredth.

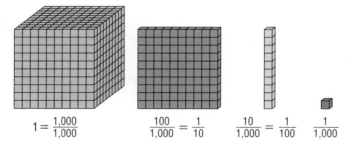

$$1 = \frac{1,000}{1,000} \qquad \frac{100}{1,000} = \frac{1}{10} \qquad \frac{10}{1,000} = \frac{1}{100} \qquad \frac{1}{1,000}$$

Together, the blocks shown above model the number 1.111. In word form, the number is read "one and one hundred eleven thousandths."

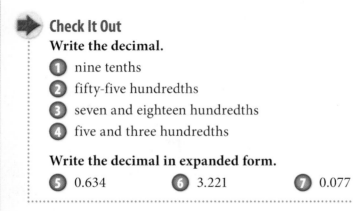

Check It Out

Write the decimal.

1. nine tenths
2. fifty-five hundredths
3. seven and eighteen hundredths
4. five and three hundredths

Write the decimal in expanded form.

5. 0.634   6. 3.221   7. 0.077

Naming Decimals Greater Than and Less Than One

Decimal numbers are based on units of ten.

Place-Value Chart

10,000	1,000	100	10	1	0.1	0.01	0.001	0.0001	0.00001
					$\frac{1}{10}$	$\frac{1}{100}$	$\frac{1}{1000}$	$\frac{1}{10,000}$	$\frac{1}{100,000}$
ten-thousands	thousands	hundreds	tens	ones	tenths	hundredths	thousandths	ten-thousandths	hundred-thousandths

You can use a place-value chart to help you name decimals greater than and less than one.

EXAMPLE **Naming Decimals Greater Than and Less Than One**

Find the value of the digits in the decimal number 45.6317.

- Values to the left of the decimal point are greater than 1.

  45 means 4 tens and 5 ones.

- Values to the right of the decimal point are less than 1. The digits 6317 appear to the right of the decimal point.

Place-Value Chart

10	1	0.1	0.01	0.001	0.0001
tens	ones	tenths	hundredths	thousandths	ten-thousandths
4	5	6	3	1	7

- The word name of the decimal is determined by the place value of the digit in last place.

  The last digit (7) is in the ten-thousandths place.

45.6317 is read as forty-five *and* six thousand three hundred seventeen ten-thousandths.

➡️ **Check It Out**

Use the place-value chart to find the value of each boldfaced digit. Then write the numbers in words.

8 5.6**3**3

9 0.0**4**5

10 6.00**7**4

11 0.0027**1**

Comparing Decimals

Zeros can be added to the right of the decimal in the following manner without changing the value of the number.

$$1.045 = 1.0450 = 1.04500 = 1.04500 \ldots$$

To compare decimals, you compare the digits in each place value.

EXAMPLE **Comparing Decimals**

Compare 18.4053 and 18.4063.

• Start at the left. Find the first place where the numbers are different.

18.40**5**3 and 18.40**6**3

The digits in the thousandths place are different.

• Compare the digits that are different.

5 < 6

• The numbers compare the same way that the digits compare.

18.4053 < 18.4063

➡️ **Check It Out**

Write <, >, or = for each □.

12 37.5 □ 37.60

13 15.336 □ 15.636

14 0.0018 □ 0.0015

Ordering Decimals

To order more than two decimals from least to greatest and vice versa, you first compare the numbers two at a time.

Order the decimals: 1.123, 0.123, 1.13 from least to greatest.

- Compare the numbers two at a time.

  $1.123 > 0.123$ $1.13 > 1.123$ $1.13 > 0.123$

- List the decimals from least to greatest.

  0.123, 1.123, 1.13

➡ Check It Out

Write in order from least to greatest.

15 4.0146, 40.146, 4.1406

16 8.073, 8.373, 8, 83.037

17 0.522, 0.552, 0.52112, 0.5512

Rounding Decimals

Rounding decimals is similar to rounding whole numbers. Round 13.046 to the nearest hundredth.

- Find the rounding place.   13.046

  hundredths

- Look at the digit to the right of the rounding place. 13.04**6**
- If it is less than 5, leave the digit in the rounding place unchanged. If it is greater than or equal to 5, increase the digit in the rounding place by 1. $6 > 5$
- Write the rounded number. 13.05

13.046 rounded to the nearest hundredth is 13.05.

All digits to the right of the rounded digit are zero. They do not change the value of the decimal and are dropped.

➡ Check It Out

Round each decimal to the nearest hundredth.

18 1.656   **19** 226.948   **20** 7.399   **21** 8.594

2.5 Exercises

Write the decimal in standard form.

1. four and twenty-six hundredths

2. five tenths

3. seven hundred fifty-six ten-thousandths

Write the decimal in expanded form.

4. seventy-six thousandths

5. seventy-five and one hundred thirty-four thousandths

Give the value of each bold digit.

6. 34.2**4**1   7. 4.3**4**61   8. 0.129**6**   9. **2**4.14

Compare. Use <, >, or = for each □.

10. 14.0990 □ 14.11

11. 13.46400 □ 13.46

12. 8.1394 □ 8.2

13. 0.664 □ 0.674

List in order from least to greatest.

14. 0.707, 0.070, 0.70, 0.777

15. 5.722, 5.272, 5.277, 5.217

16. 4.75, 0.75, 0.775, 77.5

Round each decimal to the indicated place.

17. 1.7432; tenths

18. 49.096; hundredths

19. Five girls are entered into a gymnastics competition in which the highest possible score is 10.0. On the floor routine, Rita scored 9.3, Minh 9.4, Sujey 9.9, and Sonja 9.8. What score does Aisha have to receive in order to win the competition?

20. Based on the chart below, which bank offers the savings account with the best interest rate?

Savings Banks	Interest
First Federal	7.25
Western Trust	7.125
National Savings	7.15
South Central	7.1

2·6 Decimal Operations

Adding and Subtracting Decimals

Adding and subtracting decimals is similar to adding and subtracting whole numbers.

EXAMPLE **Adding and Subtracting Decimals**

Add 3.65 + 0.5 + 22.45.

$$\begin{array}{r} 3.65 \\ 0.5 \\ + 22.45 \\ \hline \end{array}$$

• Line up the decimal points.

$$\begin{array}{r} \overset{1}{} \\ 3.65 \\ 0.5 \\ + 22.45 \\ \hline 0 \end{array}$$

• Add or subtract the place farthest right. Regroup, if necessary.

$$\begin{array}{r} \overset{1\,1}{} \\ 3.65 \\ 0.5 \\ + 22.45 \\ \hline 60 \end{array}$$

• Add or subtract the next place left. Regroup, if necessary.

$$\begin{array}{r} 3.65 \\ 0.5 \\ + 22.45 \\ \hline 26.60 \end{array}$$

• Continue through the whole numbers. Place the decimal point in the result. When you add or subtract decimals, the decimal point in the result remains in line with the decimals above it.

Check It Out

Solve.

1 18.68 + 47.30 + 22.9

2 16.8 + 5.99 + 39.126

3 6.77 − 0.64

4 47.026 − 0.743

Estimating Decimal Sums and Differences

One way that you can estimate decimal sums and differences is to use compatible numbers. Remember that compatible numbers are numbers that are close to the real numbers in the problem but easier to work with mentally.

EXAMPLE Estimating Decimal Sums and Differences

Estimate the sum of 1.344 + 8.744.

1.344 $\rightarrow$ 1	• Replace the numbers with compatible
8.744 $\rightarrow$ 9	numbers.
1 + 9 = 10	• Add the numbers.

So, 1.344 + 8.744 is about 10.

Estimate the difference of 18.572 − 7.231.

18.572 $\rightarrow$ 18	• Replace the numbers with compatible
7.231 $\rightarrow$ 7	numbers.
18 − 7 = 11	• Subtract the compatible numbers.

So, 18.572 − 7.231 is about 11.

Check It Out

Estimate each sum or difference.

5 7.64 + 4.33

6 12.4 − 8.3

7 19.144 − 4.66

8 2.66 + 3.14 + 6.54

Multiplying Decimals

Multiplying decimals is much the same as multiplying whole numbers. You can model the multiplication of decimals with a 10-by-10 grid. Each tiny square is equal to one hundredth.

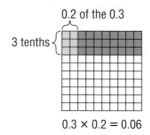

0.2 of the 0.3

3 tenths

$0.3 \times 0.2 = 0.06$

EXAMPLE Multiplying Decimals

Multiply 24.5×0.07.

- Multiply as with whole numbers.

$$
\begin{array}{r} 24.5 \\ \times\ 0.07 \\ \hline \end{array}
\qquad
\begin{array}{r} 245 \\ \times\ 7 \\ \hline 1715 \end{array}
$$

- Count the number of decimal places in the factors.

$$
\begin{array}{r} 24.5 \\ \times\ 0.07 \\ \hline 1715 \end{array}
\ \longrightarrow\
\begin{array}{l} 1 \text{ decimal place} \\ 2 \text{ decimal places} \\ 1 + 2 = 3 \text{ decimal places in the answer} \end{array}
$$

- Place the decimal point in the product. Because there are three decimal places in the factors, the decimal is placed three digits from the right in the product.

$$
\begin{array}{r} 24.5 \\ \times\ 0.07 \\ \hline 1.715 \end{array}
\ \longrightarrow\
\begin{array}{l} 1 \text{ decimal place} \\ 2 \text{ decimal places} \\ 3 \text{ decimal places} \end{array}
$$

So, $24.5 \times 0.07 = 1.715$.

Check It Out

9 2.8×1.68

10 33.566×3.4

Multiplying Decimals with Zeros in the Product

Sometimes when you are multiplying decimals, you need to add zeros in the product.

EXAMPLE Multiplying with Zeros in the Product

Multiply 0.9 × 0.0456.

- Multiply as with whole numbers. Count the decimal places in the factors to find the number of places needed in the product.

$$\begin{array}{r} 0.0456 \\ \times\ \ 0.9 \\ \hline \end{array} \qquad \begin{array}{r} 456 \\ \times\ \ 9 \\ \hline 4104 \end{array} \qquad \text{You need 5 decimal places.}$$

- Add zeros in the product, as necessary. Because 5 decimal places are needed in the product, write one zero to the left of the 4 as a place holder.

So, 0.0456 × 0.9 = 0.04104.

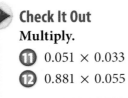 **Check It Out**

Multiply.

⑪ 0.051 × 0.033

⑫ 0.881 × 0.055

Estimating Decimal Products

To estimate decimal products, you can replace given numbers with compatible numbers. Estimate the product of 37.3 × 48.5.

- Replace the factors with compatible numbers.

  37.3 ⟶ 40

  48.5 ⟶ 50

- Multiply mentally.

  40 × 50 = 2,000

Check It Out

Estimate each product, using compatible numbers.

⑬ 34.84 × 6.6

⑭ 43.87 × 10.63

Dividing Decimals

Dividing decimals is similar to dividing whole numbers. You can use a model to help you understand how to divide decimals. For example, $0.8 \div 0.2$ means how many groups of 0.2 are in 0.8? There are 4 groups of 0.2 in 0.8, so $0.8 \div 0.2 = 4$.

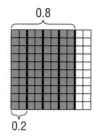

EXAMPLE Dividing Decimals

Divide $0.592 \div 1.6$.

$1.6 \times 10 = 16$

- Multiply the divisor by a power of ten so that it is a whole number.

$0.592 \times 10 = 5.92$

- Multiply the dividend by the same power of ten.

$$\begin{array}{r} 0.37 \\ 16\overline{)5.92} \\ -48 \\ \hline 112 \\ -112 \\ \hline 0 \end{array}$$

- Divide. Place the decimal point in the quotient.

So, $0.592 \div 1.6 = 0.37$.

Check It Out
Divide.

15 $10.5 \div 2.1$

16 $0.0936 \div 0.02$

17 $3.024 \div 0.06$

18 $3.68 \div 0.08$

Rounding Decimal Quotients

You can use a calculator to divide decimals. Then you can follow these steps to round the quotient.

EXAMPLE **Using a Calculator to Round**

Divide 8.3 ÷ 3.6. Round to the nearest hundredth.

- Use your calculator to divide.

  8.3 $\boxed{\div}$ 3.6 $\boxed{=}$ 2.3055555

- To round the quotient, look at one place to the right of the rounding place.

  2.30**5**

- If the digit to the right of the rounding place is 5 or above, round up. If the digit to the right of the rounding place is less than 5, the digit to be rounded stays the same.

  5 = 5, so 2.305555556 rounded to the nearest hundredth is 2.31.

So, 8.3 ÷ 3.6 ≈ 2.31.

Check It Out

Use a calculator to find each quotient. Round to the nearest hundredth.

19 0.509 ÷ 0.7

20 0.1438 ÷ 0.56

21 0.2817 ÷ 0.47

2·6 Exercises

Estimate each sum or difference.

1. $4.64 + 2.44$

2. $7.09 + 4.7$

3. $6.666 + 0.34$

4. $4.976 + 3.224$

5. $12.86 - 7.0064$

Add.

6. $224.2 + 3.82$

7. $55.12 + 11.65$

8. $10.84 + 174.99$

9. $8.0217 + 0.71$

10. $1.9 + 6 + 2.5433$

Subtract.

11. $24 - 10.698$

12. $32.034 - 0.649$

13. $487.1 - 3.64$

14. $53.44 - 17.844$

15. $11.66 - 4.0032$

Multiply.

16. 0.5×5.533

17. 11.5×23.33

18. 0.13×0.03

19. 39.12×0.5494

20. 0.47×0.81

Divide.

21. $273.5 \div 20.25$

22. $29.3 \div 0.4$

23. $76.5 \div 25.5.$

24. $38.13 \div 8.2$

Use a calculator to divide. Round the quotient to the nearest hundredth.

25. $583.5 \div 13.2$

26. $798.46 \div 92.3$

27. $56.22 \div 0.28$

28. $0.226 \div 0.365$

29. The school's record in the field day relay race was 45.78 seconds. This year the record was broken by 0.19 second. What was the new record time this year?

30. Arthur delivers pizzas for $4.75 an hour. Last week he worked 43 hours. How much did he earn?

2·7 Meaning of Percent

Naming Percents

Percent is a **ratio** that compares a number with 100. Percent means *per hundred* and is represented by the symbol %.

You can use graph paper to model percents. There are 100 squares in a 10-by-10 grid. So, the grid can be used to represent 100%. Because percent means how many out of 100, it is easy to tell what percent of the 100-square grid is shaded.

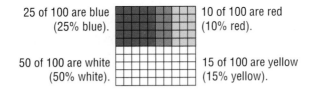

25 of 100 are blue
(25% blue).

10 of 100 are red
(10% red).

50 of 100 are white
(50% white).

15 of 100 are yellow
(15% yellow).

Check It Out

Give the percents for the number of squares that are shaded and the number of squares that are not shaded.

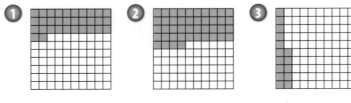

Understanding the Meaning of Percent

Any percent ratio can be expressed in three ways. You can write the ratio as a fraction, a decimal, and a percent.

A quarter is 25% of $1.00. You can express a quarter as 25¢, $0.25, $\frac{1}{4}$ of a dollar, $\frac{25}{100}$, and 25%.

You can build what you know about percents based on these few **benchmarks**. You can use these benchmarks to help you estimate percents.

EXAMPLE Estimating Percents

Estimate 47% of 60.

47% is close to 50%.

$50\% = \dfrac{1}{2}$

$\dfrac{1}{2}$ of 60 is 30.

- Choose a benchmark, or combination of benchmarks, close to the target percent.
- Find the fraction or decimal equivalent to the benchmark percent.
- Use the benchmark equivalent to estimate the percent.

So, 47% of 60 is about 30.

➡️ **Check It Out**

Use fractional benchmarks to estimate the percents.

4️⃣ 34% of 70

5️⃣ 45% of 80

6️⃣ 67% of 95

7️⃣ 85% of 32

Using Mental Math to Estimate Percents

You can use fractional or decimal benchmarks in real-world situations to help you quickly estimate the percent of something, such as calculating a tip at a restaurant.

EXAMPLE Using Mental Math to Estimate Percent

Estimate a 20% tip for a bill of $15.40.

$15.40 rounds to $15.00.	• Round to a compatible number.
20% = 0.20	• Think of the percent as a benchmark.
0.20×15.00	• Multiply mentally.
$= (0.10 \times 15.00) \times 2$	
$= (1.5) \times 2 = \$3.00$	

The tip is about $3.00.

Estimate a 15% tip for a bill of $47.75.

$47.75 rounds to $50.00.	• Round to a compatible number.
15% = 0.15	• Think of the percent as a benchmark.
or 0.10 + 0.05 (half of 0.10)	
$0.10 \times 50 = 5$	• Multiply mentally.
and $\frac{1}{2} \times 5 = 2.5$	
$5 + 2.5 = 7.5$	• Add the two benchmark parts.

The tip is about $7.50.

Check It Out

Use mental math to estimate each percent.

8 10% of $14.55

9 23% of $16

10 17% of $110

2·7 Exercises

Write the percent for the amount that is shaded and for the amount that is not shaded.

1. 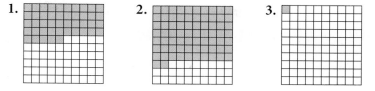 2. 3.

Write each ratio as a fraction, a decimal, and a percent.

4. 8 to 100

5. 23 to 100

6. 59 to 100

Use fractional benchmarks to estimate the percents of each number.

7. 27% of 60

8. 49% of 300

9. 11% of 75

10. 74% of 80

Use mental math to estimate each percent.

11. 15% of $45

12. 20% of $29

13. 10% of $79

14. 25% of $69

15. 6% of $35

2·8 Using and Finding Percents

Finding a Percent of a Number

There are several ways that you can find the percent of a number. To find the percent of a number, you must first change the percent to a decimal or a fraction. Sometimes it is easier to change to a decimal representation and other times to a fractional one.

To find 50% of 80, you can use either the fraction method or the decimal method.

EXAMPLE Finding the Percent of a Number: Two Methods

Find 50% of 80.

Decimal Method
- Change the percent to an equivalent decimal.

  $50\% = 0.5$
- Multiply.

  $0.5 \times 80 = 40$

Fraction Method
- Change the percent to a fraction in lowest terms.

  $50\% = \dfrac{50}{100} = \dfrac{1}{2}$
- Multiply.

  $\dfrac{1}{2} \times 80 = 40$

So, 50% of 80 is equal to 40.

Check It Out

Give the percent of each number.

① 55% of 35

② 94% of 600

③ 22% of 55

④ 71% of 36

Using Part to Whole to Find Percent

Remember that a *ratio* is a comparison of two numbers. A **proportion** is a statement that two ratios are equal (p. 238).

You can use a proportion to solve percent problems. One ratio of the proportion compares a part of the quantity to the whole. The other ratio is the percent written as a fraction.

$$\frac{\text{part}}{\text{whole}} = \frac{n}{100} \text{ \} percent}$$

EXAMPLE **Finding the Percent**

What percent of 30 is 6?
(30 is the *whole*; 6 is the *part*.)

$\dfrac{\text{part}}{\text{whole}} = \dfrac{\text{parts of}}{100}$ • Set up a proportion, using this form.

$\dfrac{6}{30} = \dfrac{n}{100}$ (*n* stands for the missing number you want to find.)

$100 \times 6 = 30 \times n$ • Show the cross products of the proportion.

$600 = 30n$ • Find the products.

$\dfrac{600}{30} = \dfrac{30n}{30}$ • Divide both sides of the equation by 30.

$n = 20$

So, 6 is 20% of 30.

Check It Out

Solve.

5 What percent of 240 is 60?

6 What percent of 500 is 75?

7 What percent of 60 is 3?

8 What percent of 44 is 66?

EXAMPLE Finding the Whole

60 is 48% of what number? (The phrase *what number* refers to the *whole*.)

$$\frac{\text{part}}{\text{whole}} = \frac{\text{parts of}}{100}$$ • Set up a percent proportion using this form.

$$\frac{60}{n} = \frac{48}{100}$$

$60 \times 100 = 48 \times n$ • Show the cross products of the proportion.

$6{,}000 = 48n$ • Find the products.

$$\frac{6{,}000}{48} = \frac{48n}{48}$$ • Divide both sides of the equation by 48.

$n = 125$

So, 60 is 48% of 125.

Check It Out

Solve. Round the quotient to the nearest hundredth.

9 54 is 50% of what number?

10 16 is 80% of what number?

11 35 is 150% of what number?

12 74 is 8% of what number?

Using the Proportion Method

You can also use **proportions** (p. 238) to help you find the percent of a number.

EXAMPLE **Use a Proportion to Find a Percent of a Number**

Forest works in a skateboard store. He receives a commission (part of the sales) of 15% on his sales. Last month he sold $1,400 worth of skateboards, helmets, knee pads, and elbow pads. What was his commission?

$$\frac{\text{part}}{\text{whole}} = \frac{\text{parts of}}{100}$$

- Use a proportion, a statement of equal ratios, to find the percent of his sales.

$$\frac{\text{commission (unknown)}}{\text{sales (\$1,400)}} = \frac{15}{100}$$

- Identify the given items before trying to find the unknown.

$$\frac{n}{1,400} = \frac{15}{100}$$

- Call the unknown n. Set up the proportion.

$$n \times 100 = 15 \times 1,400$$

- Cross multiply.

$$100n = 21,000$$

$$\frac{100n}{100} = \frac{21,000}{100}$$

- Divide both sides by 100.

$$n = 210$$

Forest received a commission of $210.

Check It Out

Use a proportion to find the percent of each number.

13 56% of 65

14 67% of 139

15 12% of 93

16 49% of 400

Estimating a Percent

You can use what you know about compatible numbers and simple fractions to estimate a percent of a number. You can use the table to help you estimate the percent of a number.

Percent	1%	5%	10%	20%	25%	$33\frac{1}{3}$%	50%	$66\frac{2}{3}$%	75%	100%
Fraction	$\frac{1}{100}$	$\frac{1}{20}$	$\frac{1}{10}$	$\frac{1}{5}$	$\frac{1}{4}$	$\frac{1}{3}$	$\frac{1}{2}$	$\frac{2}{3}$	$\frac{3}{4}$	1

EXAMPLE   Estimating a Percent of a Number

Estimate 17% of 46.

17% is about 20%.
- Find the percent that is closest to the percent you are asked to find.

20% is equivalent to $\frac{1}{5}$.
- Find the fractional equivalent for the percent.

46 is about 50.
- Find a compatible number for the number you are asked to find the percent of.

$\frac{1}{5}$ of 50 is 10.
- Use the fraction to find the percent.

So, 17% of 46 is about 10.

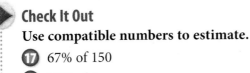 **Check It Out**

Use compatible numbers to estimate.

17 67% of 150

18 35% of 6

19 27% of 54

20 32% of 89

Honesty Pays

David Hacker, a cabdriver, found a wallet in the back seat of the cab that contained $25,000—over half a year's salary for him.

The owner's name was in the wallet, and Hacker remembered where he had dropped him off. He went straight to the hotel and found the man.

The owner, a businessman, had already realized that he had lost his wallet and figured that he would never see it again. He didn't believe that anyone would be that honest! On the spot, he handed the cabdriver fifty $100 bills.

What percent of the money did Hacker receive as a reward? See **HotSolutions** for answer.

2·8 Exercises

Find the percent of each number.

  1. 7% of 34

  2. 34% of 135

  3. 85% of 73

  4. 3% of 12.4

  5. 12% of 942

  6. 94% of 200

Solve.

  7. What percent of 500 is 35?

  8. What percent of 84 is 147?

  9. 52 is what percent of 78?

  10. What percent of 126 is 42?

  11. 70 is what percent of 1,000?

  12. 84 is what percent of 252?

Solve. Round to the nearest hundredth.

13. 38% of what number is 28?

14. 23% of what number is 13?

15. 97% of what number is 22?

16. 65% of what number is 34.2?

Estimate the percent of each number.

17. 12% of 72

18. 29% of 185

19. 79% of 65

20. 8% of 311

21. 4% of 19

22. 68% of 11

2·9 Fraction, Decimal, and Percent Relationships

Percents and Fractions

Percents and fractions both describe a ratio out of 100. The chart below shows the relationship between percents and fractions.

Percent	Fraction
50 out of 100 = 50%	$\frac{50}{100} = \frac{1}{2}$
$33\frac{1}{3}$ out of 100 = $33\frac{1}{3}$%	$\frac{33.\overline{3}}{100} = \frac{1}{3}$
25 out of 100 = 25%	$\frac{25}{100} = \frac{1}{4}$
20 out of 100 = 20%	$\frac{20}{100} = \frac{1}{5}$
10 out of 100 = 10%	$\frac{10}{100} = \frac{1}{10}$
1 out of 100 = 1%	$\frac{1}{100}$
$66\frac{2}{3}$ out of 100 = $66\frac{2}{3}$%	$\frac{66.\overline{6}}{100} = \frac{2}{3}$
75 out of 100 = 75%	$\frac{75}{100} = \frac{3}{4}$

You can write fractions as percents and percents as fractions.

EXAMPLE **Converting a Fraction to a Percent**

Use a proportion to express $\frac{2}{5}$ as a percent.

$\frac{2}{5} = \frac{n}{100}$ • Set up a proportion.

$5n = 2 \times 100$ • Solve the proportion.

$n = 40$

$\frac{2}{5} = 40\%$ • Express as a percent.

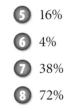

Check It Out

Change each fraction to a percent. Round to the nearest whole percent.

1 $\frac{11}{20}$

2 $\frac{4}{10}$

3 $\frac{6}{8}$

4 $\frac{3}{7}$

Changing Percents to Fractions

To change from a percent to a fraction, write the percent as the numerator of a fraction with a denominator of 100, and express in simplest form.

EXAMPLE	Changing Percents to Fractions

Express 45% as a fraction.

- Change the percent directly to a fraction with a denominator of 100. The number of the percent becomes the numerator of the fraction.

  $45\% = \frac{45}{100}$

- Express the fraction in simplest form.

  $\frac{45}{100} = \frac{9}{20}$

So, 45% expressed as a fraction in simplest form is $\frac{9}{20}$.

Check It Out

Convert each percent to a fraction in simplest form.

5 16%

6 4%

7 38%

8 72%

Changing Mixed Number Percents to Fractions

Changing a mixed number percent to a fraction is similar to changing percents to fractions.

EXAMPLE **Changing Mixed Number Percents to Fractions**

Write $15\frac{1}{4}\%$ as a fraction.

$15\frac{1}{4}\% = \frac{61}{4}\%$ — • Change the mixed number to an improper fraction.

$\frac{61}{4} \times \frac{1}{100} = \frac{61}{400}$ — • Multiply the percent by $\frac{1}{100}$.

$15\frac{1}{4}\% = \frac{61}{400}$ — • Simplify, if possible.

So, $15\frac{1}{4}\%$ written as a fraction is $\frac{61}{400}$.

Check It Out

Convert each mixed number percent to a fraction expressed in simplest form.

9 $24\frac{1}{2}\%$

10 $16\frac{3}{4}\%$

11 $121\frac{1}{8}\%$

Percents and Decimals

Percents can be expressed as decimals, and decimals can be expressed as percents. *Percent* means part of a hundred or hundredths.

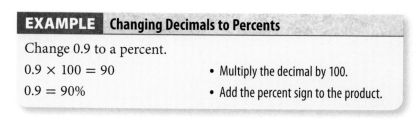

EXAMPLE **Changing Decimals to Percents**

Change 0.9 to a percent.

$0.9 \times 100 = 90$ — • Multiply the decimal by 100.

$0.9 = 90\%$ — • Add the percent sign to the product.

A Shortcut for Changing Decimals to Percents

Change 0.9 to a percent.

- Move the decimal point two places to the right. Add zeros, if necessary.

    0.9 ⟶ 0.90.

- Add the percent sign.

    90%

So, 0.9 = 90%.

➡️ **Check It Out**

Write each decimal as a percent.

12 0.45

13 0.606

14 0.019

15 2.5

Changing Percents to Decimals

Because *percent* means part of a hundred, percents can be converted directly to decimals.

EXAMPLE **Changing Percents to Decimals**

Change 6% to a decimal.

$6\% = \frac{6}{100}$ • Express the percent as a fraction with 100 as the denominator.

$6 \div 100 = 0.06$ • Change the fraction to a decimal by dividing the numerator by the denominator.

So, 6% = 0.06.

A Shortcut for Changing Percents to Decimals

Change 6% to a decimal.

- Move the decimal point two places to the left.

    6% → ⌣.6.

- Add zeros, if necessary.

    6% = 0.06

Check It Out

Express each percent as a decimal.

16 54%

17 190%

18 4%

19 29%

Fractions and Decimals

Fractions can be written as either **terminating** or **repeating** **decimals**.

Fractions	Decimals	Terminating or Repeating
$\frac{1}{2}$	0.5	terminating
$\frac{1}{3}$	0.3333333 . . .	repeating
$\frac{1}{6}$	0.166666 . . .	repeating
$\frac{2}{3}$	0.666666 . . .	repeating
$\frac{3}{5}$	0.6	terminating

EXAMPLE Changing Fractions to Decimals

Write $\frac{2}{5}$ as a decimal.

$2 \div 5 = 0.4$ • Divide the numerator by the denominator.

The remainder is zero.

0.4 is a terminating decimal.

Write $\frac{2}{3}$ as a decimal.

$2 \div 3 = 0.666666\ldots$ • Divide the numerator by the denominator.

The decimal is a repeating decimal.

$0.6\overline{6}$ or $0.\overline{6}$ • Place a bar over the digit that repeats.

So, $\frac{2}{3} = 0.6\overline{6}$. It is a repeating decimal.

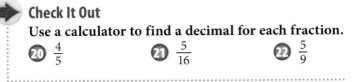 **Check It Out**

Use a calculator to find a decimal for each fraction.

20 $\frac{4}{5}$ **21** $\frac{5}{16}$ **22** $\frac{5}{9}$

EXAMPLE Changing Decimals to Fractions

Write the decimal 0.24 as a fraction.

$0.24 = \frac{24}{100}$ • Write the decimal as a fraction.

$\frac{24}{100} = \frac{24 \div 4}{100 \div 4} = \frac{6}{25}$ • Express the fraction in simplest form.

So, $0.24 = \frac{6}{25}$.

Check It Out

Write each decimal as a fraction.

23 0.225   **24** 0.5375   **25** 0.36

2·9 Exercises

Change each fraction to a percent.

1. $\frac{1}{5}$

2. $\frac{3}{25}$

3. $\frac{1}{100}$

4. $\frac{13}{50}$

Change each percent to a fraction in simplest form.

5. 28%

6. 64%

7. 125%

8. 87%

Write each decimal as a percent.

9. 0.9

10. 0.27

11. 0.114

12. 0.55

13. 3.7

Write each percent as a decimal.

14. 38%

15. 13.6%

16. 19%

17. 5%

18. 43.2%

Change each fraction to a decimal. Use a bar to show repeating digits.

19. $\frac{1}{5}$

20. $\frac{2}{9}$

21. $\frac{3}{16}$

22. $\frac{4}{9}$

23. $\frac{9}{10}$

Write each decimal as a fraction in simplest form.

24. 0.05

25. 0.005

26. 10.3

27. 0.875

28. 0.6

29. Bargain Barn is offering CD players at 50% off the regular price of $149.95. Larry's Lowest is offering CD players at $\frac{1}{3}$ off the regular price of $119.95. Which store has the better buy?

30. One survey at Franklin Middle School said that 24% of the sixth-grade students named basketball as their favorite sport. Another survey said that $\frac{6}{25}$ of the sixth-grade students named basketball as their favorite sport. Could both surveys be correct? Explain.

Fractions, Decimals, and Percents

What have you learned?

You can use the problems and the list of words that follow to see what you learned in this chapter. You can find out more about a particular problem or word by referring to the topic number (*for example,* Lesson 2·2).

Problem Set

1. Miguel bought 0.8 kg of grapes at $0.55 a kilogram and 15 grapefruits at $0.69 each. How much did he spend?
   (Lesson 2·6)

2. Which fraction is equivalent to $\frac{16}{24}$? (Lesson 2·1)

   A. $\frac{2}{3}$ **B.** $\frac{8}{20}$ **C.** $\frac{4}{5}$ **D.** $\frac{6}{4}$

3. Which fraction is greater, $\frac{1}{16}$ or $\frac{2}{19}$? (Lesson 2·2)

Add or subtract. Write your answers in simplest form. (Lesson 2·3)

4. $\frac{3}{5} + \frac{5}{9}$

5. $4\frac{1}{7} - 2\frac{3}{4}$

6. $6 - 1\frac{2}{3}$

7. $7\frac{1}{9} + 2\frac{7}{8}$

8. Write the improper fraction $\frac{16}{5}$ as a mixed number. (Lesson 2·1)

Multiply or divide. Write your answers in simplest form.

(Lesson 2·4)

9. $\frac{4}{7} \times \frac{6}{7}$

10. $\frac{2}{3} \div 7\frac{1}{4}$

11. $4\frac{1}{4} \times \frac{3}{4}$

12. $5\frac{1}{4} \div 2\frac{1}{5}$

13. Give the place value of 5 in 432.159. (Lesson 2·5)

14. Write in expanded form: 4.613. (Lesson 2·5)

15. Write as a decimal: three hundred and sixty-six thousandths.
    (Lesson 2·5)

16. Write the following numbers in order from least to greatest: 0.660, 0.060, 0.066, 0.606. (Lesson 2·5)

Find each answer as indicated. (Lesson 2·6)

17. $12.344 + 2.89$

18. $14.66 - 0.487$

19. 34.89×0.0076

20. $0.86 \div 0.22$

Use a calculator to answer Exercises 21 and 22. Round to the nearest tenth. (Lesson 2·8)

21. What is 53% of 244?

22. Find 154% of 50.

Write each decimal as a percent. (Lesson 2·9)

23. 0.65

24. 0.05

Write each fraction as a percent. (Lesson 2·9)

25. $\dfrac{3}{8}$

26. $\dfrac{7}{20}$

Write each percent as a fraction in simplest form. (Lesson 2·9)

27. 36%

28. 248%

HotWords

Write definitions for the following words.

benchmark (Lesson 2·7)

common denominator
   (Lesson 2·2)

compatible numbers
   (Lesson 2·3)

cross product (Lesson 2·1)

denominator (Lesson 2·1)

equivalent (Lesson 2·1)

equivalent fractions
   (Lesson 2·1)

estimate (Lesson 2·3)

factor (Lesson 2·4)

fraction (Lesson 2·1)

greatest common factor
   (Lesson 2·1)

improper fraction (Lesson 2·1)

inverse operations (Lesson 2·4)

least common multiple
   (Lesson 2·2)

mixed number (Lesson 2·1)

numerator (Lesson 2·1)

percent (Lesson 2·7)

place value (Lesson 2·5)

product (Lesson 2·4)

proportion (Lesson 2·8)

ratio (Lesson 2·7)

reciprocal (Lesson 2·4)

repeating decimal (Lesson 2·9)

terminating decimal
   (Lesson 2·9)

whole number (Lesson 2·1)

HotTopic 3
Powers and Roots

What do you know?

You can use the problems and the list of words that follow to see what you already know about this chapter. The answers to the problems are in **HotSolutions** at the back of the book, and the definitions of the words are in **HotWords** at the front of the book. You can find out more about a particular problem or word by referring to the topic number (*for example,* Lesson 3·2).

Problem Set

Write each product using an exponent. (Lesson 3·1)

1. $7 \times 7 \times 7 \times 7 \times 7$
2. $n \times n \times n \times n \times n \times n \times n \times n$
3. $4 \times 4 \times 4$
4. $x \times x$
5. $3 \times 3 \times 3 \times 3$

Evaluate each square. (Lesson 3·1)

6. 2^2 7. 5^2 8. 10^2
9. 7^2 10. 12^2

Evaluate each cube. (Lesson 3·1)

11. 2^3 12. 4^3 13. 10^3
14. 7^3 15. 1^3

Evaluate each power of 10. (Lesson 3·1)

16. 10^2 17. 10^6 18. 10^{10}
19. 10^7 20. 10^1

Evaluate each square root. (Lesson 3·2)

21. $\sqrt{9}$

22. $\sqrt{25}$

23. $\sqrt{144}$

24. $\sqrt{64}$

25. $\sqrt{4}$

Estimate each square root between two consecutive numbers. (Lesson 3·2)

26. $\sqrt{20}$

27. $\sqrt{45}$

28. $\sqrt{5}$

29. $\sqrt{75}$

30. $\sqrt{3}$

Estimate each square root to the nearest thousandth. (Lesson 3·2)

31. $\sqrt{5}$

32. $\sqrt{20}$

33. $\sqrt{50}$

34. $\sqrt{83}$

35. $\sqrt{53}$

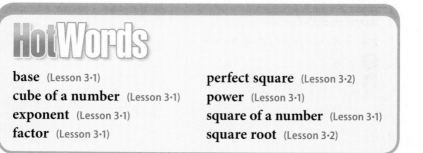

HotWords

base (Lesson 3·1)

cube of a number (Lesson 3·1)

exponent (Lesson 3·1)

factor (Lesson 3·1)

perfect square (Lesson 3·2)

power (Lesson 3·1)

square of a number (Lesson 3·1)

square root (Lesson 3·2)

3·1 Powers and Exponents

Exponents

Multiplication is the shortcut for showing a repeated addition: $4 \times 6 = 4 + 4 + 4 + 4 + 4 + 4$. A shortcut for showing the repeated multiplication $4 \times 4 \times 4 \times 4 \times 4 \times 4$ is the **power** 4^6. The 4 is the factor to be multiplied, called the **base**. The 6 is the **exponent**, which tells how many times the base is to be multiplied. The expression can be read as "4 to the sixth power." An exponent is written slightly higher than the base and is usually written smaller than the base.

EXAMPLE	Writing Products Using Exponents

Write $3 \times 3 \times 3 \times 3$ as a product using an exponent.

All the factors are 3.
- Check that the same **factor** is being used in the expression.

There are 4 factors of 3.
- Count the number of times 3 is being multiplied.

3^4
- Write the product using an exponent.

So, $3 \times 3 \times 3 \times 3 = 3^4$.

Check It Out

Write each product using an exponent.

1. $8 \times 8 \times 8 \times 8$
2. $3 \times 3 \times 3 \times 3 \times 3 \times 3 \times 3$
3. $x \times x \times x \times x$
4. $y \times y \times y \times y \times y$

Evaluating the Square of a Number

When a square is made from a segment whose length is 3, the area of the square is $3 \times 3 = 3^2 = 9$.

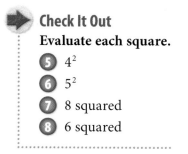

The **square of a number** means to apply the exponent 2 to a base. The square of 3 is written 3^2. To evaluate 3^2, identify 3 as the base and 2 as the exponent. Remember that the exponent tells you how many times to use the base as a factor. So, 3^2 means to use 3 as a factor 2 times:

$$3^2 = 3 \times 3 = 9$$

The expression 3^2 can be read as "3 to the second power." It can also be read as "3 squared."

EXAMPLE **Evaluating the Square of a Number**

Evaluate 7^2.

The base is 7 and the exponent is 2.
- Identify the base and the exponent.

$7^2 = 7 \times 7$
- Write the expression as a multiplication.

$7 \times 7 = 49$
- Evaluate.

So, $7^2 = 49$.

Check It Out

Evaluate each square.

5 4^2

6 5^2

7 8 squared

8 6 squared

Evaluating the Cube of a Number

To find the **cube of a number** apply the exponent 3 to a base. The cube of 2 is written 2^3. Evaluating cubes is very similar to evaluating squares. For example, if you want to evaluate 2^3, notice that 2 is the base and 3 is the exponent. Remember, the exponent tells you how many times to use the base as a factor. So, 2^3 means to use 2 as a factor 3 times:

$$2^3 = 2 \times 2 \times 2 = 8$$

The expression 2^3 can be read as "2 to the third power." It can also be read as "2 cubed."

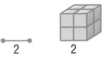

2   2

When a cube has edges of length 2, the volume of the cube is $2 \times 2 \times 2 = 2^3 = 8$.

EXAMPLE   **Evaluating the Cube of a Number**

Evaluate 2^3.

The base is 2 and the exponent is 3.
- Identify the base and the exponent.

$2^3 = 2 \times 2 \times 2$
- Write the expression as a multiplication.

$2 \times 2 \times 2 = 8$
- Evaluate.

So, $2^3 = 8$.

You can use a calculator to evaluate powers. You simply use the calculator to multiply the correct number of times, or you can use special keys (p. 327).

Check It Out

Evaluate each cube.

9 3^3 **10** 6^3 **11** 9 cubed   **12** 5 cubed

Powers of Ten

Our decimal system is based on 10. For each factor of 10, the decimal point moves one place to the right.

$3.151 \rightarrow 31.51$ $14.25 \rightarrow 1.425$ $3. \rightarrow 30$
    $\times 10$ $\times 100$ $\times 10$

When the decimal point is at the end of a number and the number is multiplied by 10, a zero is added at the end of the number.

Try to discover a pattern for the powers of 10.

Power	As a Multiplication	Result	Number of Zeros
10^2	10×10	100	2
10^4	$10 \times 10 \times 10 \times 10$	10,000	4
10^5	$10 \times 10 \times 10 \times 10 \times 10$	100,000	5
10^8	$10 \times 10 \times 10 \times 10 \times 10 \times 10 \times 10 \times 10$	100,000,000	8

Notice that the number of zeros after the 1 is the same as the power of 10. This means that, to evaluate 10^9, simply write a 1 followed by 9 zeros: 1,000,000,000.

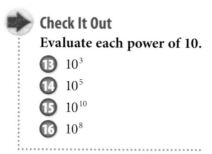

Check It Out

Evaluate each power of 10.

13 10^3

14 10^5

15 10^{10}

16 10^8

Revisiting Order of Operations

In Lesson 1·3 you learned the order of operations used to solve a problem. You learned the reminder "PEMDAS" to help you remember the order of operations.

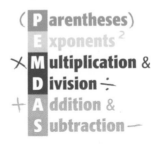

(**P** arentheses)
E xponents ^{2}
× **M** ultiplication &
D ivision ÷
+ **A** ddition &
S ubtraction —

EXAMPLE Using the Order of Operations

Simplify $2 + 2^3 \times (3^2 - 5)$.

$2 + 2^3 \times (3 \times 3 - 5)$
$2 + 2^3 \times (9 - 5)$
$2 + 2^3 \times 4$

- Simplify within the parentheses by using the order of operations, starting with the power.

$2 + 2^3 \times 4$
$2 + 2 \times 2 \times 2 \times 4$
$2 + 8 \times 4$

- Evaluate powers.

$2 + 8 \times 4$
$2 + 32$

- Multiply or divide from left to right.

$2 + 32 = 34$

- Add or subtract from left to right.

So, $2 + 2^3 \times (3^2 - 5) = 34$.

Check It Out

Find the value of each expression using the order of operations.

17 $3 \times (2 + 4^2)$

18 $(3 \times 2) + 4^2$

19 $4 \times 5^2 + 5^2$

20 $4 \times (5^2 + 5^2)$

3·1

POWERS AND EXPONENTS

3·1 Exercises

Write each product using an exponent.

1. $7 \times 7 \times 7$
2. $6 \times 6 \times 6 \times 6 \times 6 \times 6 \times 6 \times 6$
3. $y \times y \times y \times y \times y \times y$
4. $m \times m \times m \times m \times m \times m \times m \times m \times m \times m$
5. 12×12

Evaluate each square.

6. 2^2
7. 7^2
8. 10^2
9. 1 squared
10. 15 squared

Evaluate each cube.

11. 3^3
12. 8^3
13. 11^3
14. 10 cubed
15. 7 cubed

Evaluate each power of 10.

16. 10^2
17. 10^6
18. 10^{14}

19. What is the area of a square whose sides have a length of 9?
   A. 18 **B.** 36 **C.** 81 **D.** 729

20. What is the volume of a cube whose sides have a length of 5?
   A. 60 **B.** 120 **C.** 125 **D.** 150

Solve.

21. $5 + (12 - 3)$
22. $8 - 3 \times 2 + 7$
23. $5 \times 3^2 - 7$
24. $16 - 24 \div 6 \times 2$

3·2 Square Roots

Square Roots

In mathematics, certain operations are opposites of each other. That is, one operation "undoes" the other. For example, addition undoes subtraction: $3 - 2 = 1$, so $1 + 2 = 3$. Multiplication undoes division: $6 \div 3 = 2$, so $2 \times 3 = 6$. These are called *inverse operations* (p. 119).

Finding the **square root** of a number undoes the squaring of that number. You know that 3 squared $= 3^2 = 9$. The square root of 9 is the number that can be multiplied by itself to get 9, which is 3. The symbol for square root is $\sqrt{}$. Therefore, $\sqrt{9} = 3$.

EXAMPLE Finding the Square Root

Find $\sqrt{25}$.

$5 \times 5 = 25$ • Think, what number times itself makes 25?

                         • Find the square root.

So, $\sqrt{25} = 5$.

Check It Out

Find each square root.

1 $\sqrt{16}$ 2 $\sqrt{25}$ 3 $\sqrt{64}$ 4 $\sqrt{100}$

Estimating Square Roots

The table shows the first ten **perfect squares** and their square roots.

Perfect square	1	4	9	16	25	36	49	64	81	100
Square root	1	2	3	4	5	6	7	8	9	10

You can estimate the value of a square root by finding the two consecutive numbers that the square root must be between.

EXAMPLE	Estimating a Square Root

Estimate $\sqrt{40}$.

40 is between 36 and 49.	• Identify the perfect squares that 40 is between.
$\sqrt{36} = 6$ and $\sqrt{49} = 7$.	• Find the square roots of the perfect squares.
$\sqrt{40}$ is between 6 and 7.	• Estimate the square root.

So, the square root of 40 is between 6 and 7.

 Check It Out

Estimate each square root.

5 $\sqrt{20}$

6 $\sqrt{38}$

7 $\sqrt{52}$

8 $\sqrt{29}$

Better Estimates of Square Roots

If you want to find a better estimate for the value of a square root, you can use a calculator. Most calculators have a $\boxed{\sqrt{}}$ key for finding square roots.

On some calculators, the $\sqrt{}$ function is not shown on a key, but can be found above the $\boxed{x^2}$ key on the calculator's keypad. If this is true for your calculator, then you should also find a key that has either $\boxed{\text{INV}}$ or $\boxed{\text{2nd}}$ on it. To use the $\sqrt{}$ function, press $\boxed{\text{INV}}$ or $\boxed{\text{2nd}}$, and then press the key that has $\sqrt{}$ above it.

Square root key

Press the 2nd key. Then press the key with $\sqrt{}$ as the 2nd function.

| EXAMPLE | Estimating the Square Root of a Number Using a Calculator |

Use a calculator to estimate $\sqrt{42}$.

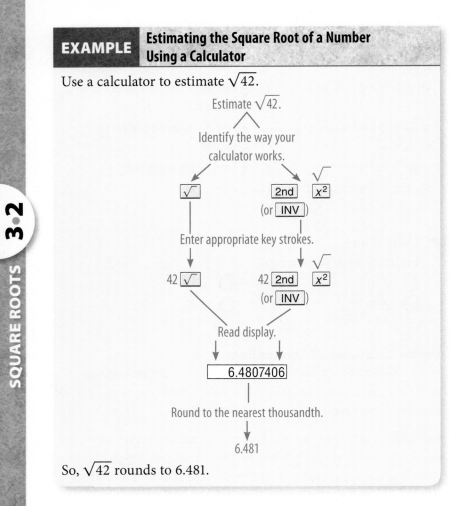

Estimate $\sqrt{42}$.

Identify the way your calculator works.

$\sqrt{}$ 2nd $\sqrt{x^2}$
(or INV)

Enter appropriate key strokes.

42 $\sqrt{}$ 42 2nd $\sqrt{x^2}$
(or INV)

Read display.

| 6.4807406 |

Round to the nearest thousandth.

6.481

So, $\sqrt{42}$ rounds to 6.481.

Check It Out

Estimate each square root to the nearest thousandth.

9 $\sqrt{2}$

10 $\sqrt{28}$

11 $\sqrt{52}$

12 $\sqrt{85}$

3·2 Exercises

Find each square root.

1. $\sqrt{9}$
2. $\sqrt{49}$
3. $\sqrt{121}$
4. $\sqrt{4}$
5. $\sqrt{144}$

6. $\sqrt{30}$ is between which two numbers?
   A. 3 and 4
   B. 5 and 6
   C. 29 and 31
   D. None of these

7. $\sqrt{72}$ is between which two numbers?
   A. 4 and 5
   B. 8 and 9
   C. 9 and 10
   D. 71 and 73

8. $\sqrt{10}$ is between which two consecutive numbers?
9. $\sqrt{41}$ is between which two consecutive numbers?
10. $\sqrt{105}$ is between which two consecutive numbers?

Estimate each square root to the nearest thousandth.

11. $\sqrt{3}$
12. $\sqrt{15}$
13. $\sqrt{50}$
14. $\sqrt{77}$
15. $\sqrt{108}$

Powers and Roots

What have you learned?

You can use the problems and the list of words that follow to see what you learned in this chapter. You can find out more about a particular problem or word by referring to the topic number (*for example*, Lesson 3·2).

Problem Set

Write each product using an exponent. (Lesson 3·1)

1. $5 \times 5 \times 5 \times 5 \times 5 \times 5 \times 5 \times 5$

2. $m \times m \times m \times m$

3. 9×9

4. $y \times y \times y \times y \times y \times y \times y \times y \times y \times y$

5. $45 \times 45 \times 45 \times 45$

Evaluate each square. (Lesson 3·1)

6. 3^2 **7.** 6^2 **8.** 12^2 **9.** 8^2 **10.** 15^2

Evaluate each cube. (Lesson 3·1)

11. 3^3 **12.** 6^3 **13.** 1^3 **14.** 8^3 **15.** 2^3

Evaluate each power of 10. (Lesson 3·1)

16. 10^3

17. 10^5

18. 10^8

19. 10^{13}

20. 10^1

Solve. (Lesson 3·1)

21. $14 + 3(7 - 2)$

22. $6 + 8 \div 2 + 2(3 - 1)$

Evaluate each square root. (Lesson 3·2)

23. $\sqrt{4}$

24. $\sqrt{36}$

25. $\sqrt{121}$

26. $\sqrt{81}$

27. $\sqrt{225}$

Estimate each square root between two consecutive numbers. (Lesson 3·2)

28. $\sqrt{27}$

29. $\sqrt{8}$

30. $\sqrt{109}$

31. $\sqrt{66}$

32. $\sqrt{5}$

Estimate each square root to the nearest thousandth. (Lesson 3·2)

33. $\sqrt{11}$

34. $\sqrt{43}$

35. $\sqrt{88}$

36. $\sqrt{6}$

37. $\sqrt{57}$

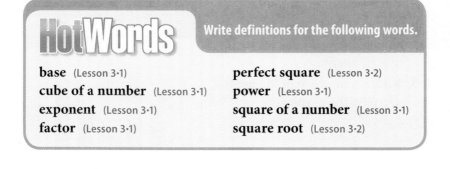

HotWords Write definitions for the following words.

base (Lesson 3·1)

cube of a number (Lesson 3·1)

exponent (Lesson 3·1)

factor (Lesson 3·1)

perfect square (Lesson 3·2)

power (Lesson 3·1)

square of a number (Lesson 3·1)

square root (Lesson 3·2)

HotTopic 4

Data, Statistics, and Probability

What do you know?

You can use the problems and the list of words that follow to see what you already know about this chapter. The answers to the problems are in **HotSolutions** at the back of the book, and the definitions of the words are in **HotWords** at the front of the book. You can find out more about a particular problem or word by referring to the topic number (*for example*, Lesson 4·2).

Problem Set

1. Jacob surveyed 20 people who were using the pool and asked them if they wanted a new pool. Is this a random sample? (Lesson 4·1)

2. Sylvia asked 40 people if they planned to vote for the school bond issue. Is the question biased or unbiased? (Lesson 4·1)

Use the following graph to answer Exercises 3–5. (Lesson 4·2)
Vanessa recorded the number of people who used the new school overpass each day.

3. What kind of graph did Vanessa make?

4. On what day did most students use the overpass?

5. Students in which grade use the overpass the most?

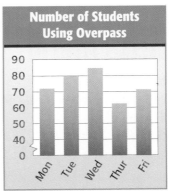

Number of Students Using Overpass

6. The weekly wages at the Ice Cream Parlor are $45, $188, $205, $98, and $155. What is the range of wages? (Lesson 4·3)

7. Find the mean and median of the wages in Exercise 6.
(Lesson 4·3)

8. Jasper grew pumpkins weighing 16, 14, 13, 15, and 76 pounds. What is the mean weight of the pumpkins? (Lesson 4·3)

9. Which data value in Exercise 8 is considered an outlier?
(Lesson 4·3)

Use the following information to answer Exercises 10 and 11.
(Lesson 4·4)

A box contains 40 tennis balls. Eighteen are green and the rest are yellow.

10. One ball is drawn. What is the probability it is yellow?
(Lesson 4·4)

11. Two balls are drawn without replacement. What is the probability they are both green? (Lesson 4·4)

HotWords

average (Lesson 4·3)

bar graph (Lesson 4·2)

circle graph (Lesson 4·2)

experimental probability
(Lesson 4·4)

leaf (Lesson 4·2)

line graph (Lesson 4·2)

mean (Lesson 4·3)

median (Lesson 4·3)

mode (Lesson 4·3)

outcome (Lesson 4·4)

outcome grid (Lesson 4·4)

outlier (Lesson 4·3)

percent (Lesson 4·2)

population (Lesson 4·1)

probability (Lesson 4·4)

probability line (Lesson 4·4)

random sample (Lesson 4·1)

range (Lesson 4·3)

sample (Lesson 4·1)

simple event (Lesson 4·4)

spinner (Lesson 4·4)

stem (Lesson 4·2)

stem-and-leaf plot (Lesson 4·2)

survey (Lesson 4·1)

table (Lesson 4·1)

tally marks (Lesson 4·1)

theoretical probability
(Lesson 4·4)

tree diagram (Lesson 4·4)

4·1 Collecting Data

Surveys

Have you ever been asked to name your favorite color? Or asked what kind of music you like? These kinds of questions are often asked in **surveys**. A group of people or objects under study is called a **population**. A small part of the population is called a **sample**.

In a survey, 150 sixth graders at Kennedy School were chosen at random and asked what kind of pets they had. The bar graph to the right shows the percent of students who named each type of pet.

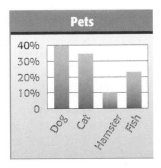

In this case, the population is all sixth graders at Kennedy School. The sample is the 150 students who were actually asked to name each type of pet they had.

In any survey:

- The population consists of the people or objects about which information is desired.
- The sample consists of the people or objects in the population that are actually studied.

Check It Out

Identify the population and the size of each sample.

1. Sixty students who were signed up for after-school sports were asked if they wanted to have the sports available during the summer.

2. Fifteen wolves on Isle Royale were tagged and let loose.

Random Samples

When you choose a sample to survey, be sure that the sample is representative of the population. The sample must be a **random sample**, where each person in the population has an equal chance of being included.

Shayna wants to find out how many of her classmates would like to have a class party at the end of the year. She picks a sample by writing the names of her classmates on cards and drawing 15 cards from a bag. She will then ask those 15 classmates whether they want to have a class party.

EXAMPLE   Determining Whether a Sample Is Random

Determine if Shayna's sample is random.
- Define the population.

  The population is the students in Shayna's class.
- Define the sample.

  The sample consists of 15 students.
- Determine if the sample is random.

Because every classmate has the same chance of being chosen, the sample is random.

Check It Out

Answer the following.

❸ How do you think you could select a random sample of your classmates?

❹ Suppose that you ask 20 people working out at a fitness center which center they prefer. Is this a random sample?

Questionnaires

It is important that the questions used in a survey are not biased. That is, the questions should not assume anything or influence the responses. The following two questionnaires are designed to find out which types of sports you like. The first questionnaire uses biased questions. The second questionnaire uses questions that are not biased.

Survey 1:
   A. Do you prefer boring sports like table tennis?
   B. Are you the adventurous type who likes to sky dive?

Survey 2:
   A. Do you like to play table tennis?
   B. Do you like to sky dive?

To develop a questionnaire:

- Decide what topic you want to ask about.
- Define a population and decide how to select a random sample from that population.
- Develop questions that are not biased.

Check It Out

Use the survey examples above to answer Exercises 5 and 6.

5 Why is **A** in Survey 1 biased?

6 Why is **B** in Survey 2 better than **B** in Survey 1?

7 Write a question that asks the same thing as the following question but is not biased: Are you a caring person who gives money to charity?

Compiling Data

Once Shayna collects the data from her classmates about a class party, she has to decide how to show the results. As she asks each classmate if they want to have a party, she uses **tally marks** to tally the answers in a table. The following **table** shows their responses.

Do You Want a Party?	Number of Students
Yes	̄|||| |||
No	||||
Don't care	|||

A table organizes data into columns and rows. To make a table to compile data:

- List the categories or questions in the first column or row.
- Tally the responses in the second column or row.

 Check It Out

Use the data in the table above to answer Exercises 8–10.

 8 How many students don't care if they have a party?

9 Which response was given by the greatest number of students?

10 If Shayna uses the survey to decide whether to have a party, what should she do? Explain.

Chilled to the Bone

Wind carries heat away from the body, increasing the cooling rate. So whenever the wind blows, you feel cooler. If you live in an area where the temperature drops greatly in winter, you know you may feel much, much colder on a blustery winter day than the temperature indicates.

Wind Speed (mi/h)	Air Temperature (°F)							
	35	30	25	20	15	10	5	0
Calm	35	30	25	20	15	10	5	0
5	31	25	19	13	7	1	−5	−11
10	27	21	15	9	3	−4	−10	−16
15	25	19	13	6	0	−7	−13	−19
20	24	17	11	4	−2	−9	−15	−22
25	23	16	9	3	−4	−11	−17	−24
30	22	15	8	1	−5	−12	−19	−26

This wind-chill table shows the effects of the cooling power of the wind in relation to temperature under calm conditions (no wind). Notice that the wind speed (in miles per hour) is correlated with the air temperature (in degrees Fahrenheit). To determine the wind-chill effect, read across and down to find the entry in the table that matches a given wind speed and temperature.

Listen to or read your local weather report each day for a week or two in the winter. Record the daily average temperature and wind speed. Use the table to determine how chilly it felt each day.

4·1 Exercises

1. One thousand registered voters were asked which political party they prefer. Identify the population and the sample. How big is the sample?

2. Livna wrote the names of 14 of her 20 classmates on slips of paper and drew five from a bag. Was the sample random?

3. LeRon asked students who ride the bus with him if they participate in clubs at school. Was the sample random?

Are the following questions biased? Explain.

4. How do you get to school?

5. Do you bring a boring lunch from home or buy a school lunch?

Write questions that ask the same thing as the following questions but are not biased.

6. Are you a caring person who recycles?

7. Do you like the uncomfortable chairs in the lunchroom?

Ms. Sandover asked her students which of the following national parks they would like to visit. The data collected is shown in the table below.

National Park	Number of Sixth Graders	Number of Seventh Graders
Yellowstone	|||| ||	|||| |||| ||
Yosemite	|||| |||| ||	||||
Olympic	|||| |||	|||| ||||
Grand Canyon	|||| ||||	|||| |||| ||||
Glacier	||||		|||| ||

8. Which park was the most popular? How many students preferred that park?

9. Did more students pick Yellowstone or Olympic?

10. How many students were surveyed?

4·2 Displaying Data

You have already seen how a frequency table is used to organize data. Here is another example: Desrie counted the number of cars that passed her school during several 15-minute periods in one day. The results are shown below.

10 14 13 12 17 18 12 18 18 11 10 13 15 18 17 10 18 10

EXAMPLE   Making a Table

Make a table to organize the data about the number of cars.

• Name the first column or row *what* you are counting.

   Label the first row *Number of Cars*.

• Tally the amounts for each category in the second row.

Number of Cars	10	11	12	13	14	15	16	17	18
Frequency	IIII	I	II	II	I	I		II	ﬀﬀ

• Count the tallies and record the number in the second row.

Number of Cars	10	11	12	13	14	15	16	17	18
Frequency	4	1	2	2	1	1	0	2	5

The most common number of cars was 18. Only once did 11, 14, and 15 cars come by during a 15-minute period.

Check It Out

Use the tables above to answer Exercise 1.

1. During how many 15-minute periods did Desrie count 10 or more cars?

2. Make a table using the data below of numbers of sponsors who were signed up by students taking part in a walk-a-thon.

   4 6 2 5 10 9 8 2 4 6 10 10 4 2 8 9 5 5 5 10 5 9

Interpret a Circle Graph

You can also use a **circle graph** to show data. In a circle graph, percentages are shown as parts of one whole. You can see how the sizes of each group compare to one another.

There are 300 students at Stratford Middle School. Toshi surveyed the entire population to find out how many of her schoolmates have brothers and sisters (siblings).

150 of the students have just one sibling.

75 students have two siblings.

30 students have three or more siblings.

45 students have no siblings.

Each number of students represents part of the whole, or a percentage of the total students.

$$150 \text{ students } = \frac{150}{300} = 50\%$$

$$75 \text{ students } = \frac{75}{300} = 25\%$$

$$45 \text{ students } = \frac{45}{300} = 15\%$$

$$30 \text{ students } = \frac{30}{300} = 10\%$$

$$= 100\%$$

Toshi made the following graph.

Students with Siblings

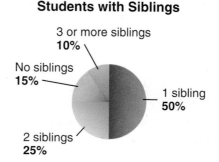

From the graph, you can see that most students in the school have siblings, and of the students with siblings, most have only one.

EXAMPLE **Interpreting a Circle Graph**

Use the circle graph to determine which sport is the most popular among students at a middle school.

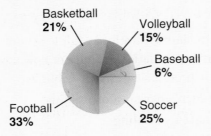

- The entire circle represents the preferences of all of the students in the school.
- The parts of the circle represent the groups of students who prefer each sport.

33% is the highest percentage shown. The highest percentage of students prefer football.

Check It Out

Use the circle graph to answer Exercises 3–5.

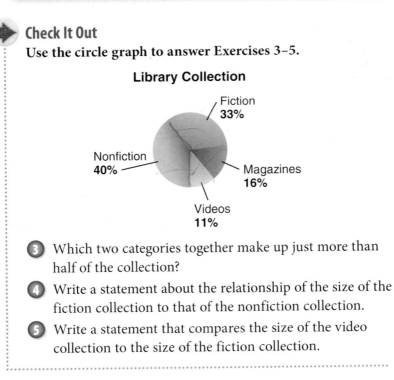

Library Collection

3 Which two categories together make up just more than half of the collection?

4 Write a statement about the relationship of the size of the fiction collection to that of the nonfiction collection.

5 Write a statement that compares the size of the video collection to the size of the fiction collection.

Create and Interpret a Line Plot

You have used tally marks to show data. Suppose you collect the following information about the number of books your classmates checked out from the library.

5 2 1 0 4 4 2 1 5 2 6 3 2 7 1 5 2 3

You can make a line plot by placing X's above a number line.

To make a line plot:

- Draw a number line showing the numbers in your data set. In this case, you would draw a number line showing the numbers 0 through 7.
- Place an X to represent each result above the number line for each number you have.
- Title the diagram.
  In this case, you could call it "Books Checked Out."

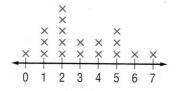

Books Checked Out

You can tell from the line plot that students checked out between 0 and 7 library books.

Check It Out

Use the "Books Checked Out" line plot to answer Exercises 6 and 7.

6 How many students checked out more than four books?

7 What is the most common number of books checked out?

8 Make a line plot to show the number of cars that went by Desrie's school (p. 178).

Interpret a Line Graph

A line graph can be used to show changes in data over time.

The following data show temperatures in degrees Fahrenheit during daytime hours on a summer day.

6:00 A.M.	8:00 A.M.	10:00 A.M.	12:00 P.M.	2:00 P.M.	4:00 P.M.	6:00 P.M.	8:00 P.M.	10:00 P.M.
58°	61°	69°	80°	84°	88°	87°	79°	77°

A line graph makes it easier to see how the temperature changes over time.

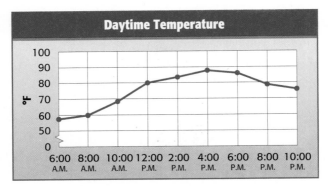

From the graph, you can see that the temperature increases throughout the morning and afternoon and then decreases again into the evening.

Check It Out

Use the graph above to answer Exercises 9–11.

9 At what time was it about 70°F?

10 Is the following statement true or false? The temperature increased more between 6:00 A.M. and 8:00 A.M. than it did between 8:00 A.M. and 10:00 A.M.

11 During which time period did the temperature drop most rapidly?

Interpret a Stem-and-Leaf Plot

The following numbers show students' scores on a math quiz.

33 27 36 18 30 24 31 33 27 32 27 35 23 40 22 34 28
31 28 28 26 31 28 32 25 29

It is hard to tell much about the scores when they are displayed
in a random list. Another way to show the information is to
make a **stem-and-leaf plot**. The following stem-and-leaf plot
shows the scores.

Math Quiz Scores

Stem	Leaf
1	8
2	2 3 4 5 6 7 7 7 8 8 8 8 9
3	0 1 1 1 2 2 3 3 4 5 6
4	0

$2 \mid 2 = 22$

Notice that the tens digits appear in the left-hand column. These
are called **stems**. Each digit on the right is called a **leaf**. By
looking at the plot, you can tell that most of the students scored
from 22 to 36 points.

Check It Out

Use this stem-and-leaf plot showing the ages of people
who came into Game-O-Mania in the mall to answer
Exercises 12–14.

Ages of Game-O-Mania Visitors

Stem	Leaf
0	7 9
1	0 2 2 4 5 8
2	1 3 4 5 6 7 8 8 8
3	0 3 4

$3 \mid 0 = 30 \ years \ old$

12 How many people came into the store?

13 Are more people in their twenties or teens?

14 Three people were the same age. What age is that?

Interpret and Create a Bar Graph

Another type of graph you can use to show data is called a **bar graph**. In this graph, either horizontal or vertical bars are used to show data. Consider the data showing the area of Rhode Island's five counties.

County	Area
Bristol	25 mi^2
Kent	170 mi^2
Newport	104 mi^2
Providence	413 mi^2
Washington	333 mi^2

Make a bar graph to show the area of Rhode Island's counties.

- Choose a vertical scale and decide what to place along the horizontal scale.

   In this case, the vertical scale can show square miles in increments of 50 square miles and the horizontal scale can show the county names.

- Above each name, draw a bar of the appropriate height.

- Write a title for the graph.

   Title this graph "Areas of Rhode Island Counties."
   Your bar graph should look like this:

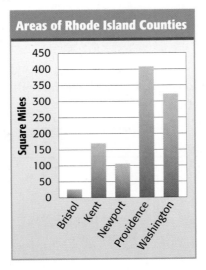

From the graph, you can see that the county with the largest area is Providence County.

Check It Out

Use the bar graph "Area of Rhode Island Counties"
(p. 184) to answer Exercises 15 and 16.

15 What is the smallest county?

16 How would the graph be different if every vertical line
represents 100 square miles instead of 50?

17 Use the following data to make a bar graph about how
students spend their time after school.

Play outdoors: 26   Do chores: 8
Talk to friends: 32   Watch TV: 18

APPLICATION **Graphic Impressions**

Both these graphs compare the maximum life span of
guppies, giant spiders, and crocodiles.

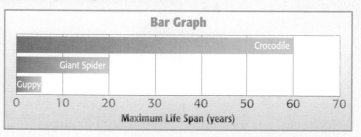

Which graph do you think more accurately portrays the
relative differences in the maximum life spans of these
three animals. What impressions does the picture graph
give you? See **HotSolutions** for answer.

4·2 Exercises

Use the data about the first words in a story to answer Exercises 1–4.

Number of Letters in the First Words in a Story

2 6 3 1 4 4 3 5 2 5 4 5 4 3 1 5 3 2 3 2

1. Make a table to show the data about first words in a story.
2. How many words were counted?
3. Make a line plot to show the data about first words.
4. Use your line plot to describe the number of letters in the words.

5. Kelsey made a circle diagram to show the favorite school subjects of 20 of her classmates.

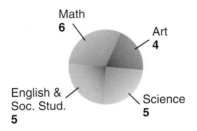

Math
6

Art
4

English &
Soc. Stud.
5

Science
5

   Which subject is the favorite of the fewest of Kelsey's classmates? Write a sentence about it.

6. The following stem-and-leaf plot shows the number of salmon traveling up a fish ladder every hour.

   Salmon Traveling up Ladder

Stem	Leaf
7	0 2 4 4 4 4 6 9
8	0 1 3 4 4 5 5 6 8 8
9	1 1 7

   8 | 3 = 83

   Draw a conclusion from the plot.

7. The sixth graders recycled 89 pounds of aluminum, the seventh graders recycled 78 pounds, and the eighth graders recycled 92 pounds. Make a bar graph to display this information.

4·3 Statistics

The *mean,* the *median,* and the *mode* are called the measures of *central tendency.* Each of these measures is calculated differently, and the one that is most useful depends on the situation.

Mei Mei asks 15 friends how much allowance they earn. She records these amounts:

$1 $1 $2 $3 $3 $3 $3 $5 $6 $6 $7 $7 $8 $8 $42

Mei Mei says her friends typically earn $3, but Navid disagrees. He says that the typical amount is $5. A third friend, Gabriel, says they are both wrong—the typical amount is $7. Each is correct because each is using a different measure to describe the central tendency.

Mean

To find the **mean**, or **average**, add the allowances together and divide by the total number of allowances being compared.

EXAMPLE **Finding the Mean**

Find the mean allowance.

- Add the amounts.

   $1 + $1 + $2 + $3 + $3 + $3 + $3 + $5 + $6 +
   $6 + $7 + $7 + $8 + $8 + $42 = $105

- Divide the total by the number of amounts.

   In this case, there are 15 allowances:
   $105 ÷ 15 = $7

The mean allowance is $7. Gabriel used the mean when he said the typical amount was $7.

Find the mean.

1. 12, 15, 63, 12, 24, 34, 23, 15
2. 84, 86, 98, 78, 82, 94
3. 132, 112, 108, 243, 400, 399, 202
4. Ryan earned money babysitting. He earned $40, $40, $51, $32, and $22. Find the mean amount he earned.

Median

Another way to describe the central tendency of numbers is to find the median. The **median** is the middle number in the data when the numbers are arranged in order from least to greatest. Let's look again at the allowances.

$1 $1 $2 $3 $3 $3 $3 $5 $6 $6 $7 $7 $8 $8 $42

EXAMPLE   Finding the Median

Find the median of the allowances.

- Arrange the data in numerical order from least to greatest or greatest to least.

  Looking at the allowances, we can see they are already arranged in order.

- Find the middle number.

  There are 15 numbers. The middle number is $5 because there are 7 numbers above $5 and 7 numbers below $5.

Navid was using the median when he described the typical allowance.

When the number of amounts is even, you can find the median by finding the mean of the two middle numbers. So, to find the median of the numbers 1, 3, 4, 3, 7, and 12, you must find the mean of the two numbers in the middle.

> **EXAMPLE** Finding the Median of an Even Number of Data
>
> - Arrange the numbers in order from least to greatest or greatest to least.
>   1, 3, 3, 4, 7, 12 or 12, 7, 4, 3, 3, 1
> - Find the mean of the two middle numbers.
>   The two middle numbers are 3 and 4:
> $(3 + 4) \div 2 = 3.5$
>
> The median is 3.5. Half the numbers are greater than 3.5 and half the numbers are less than 3.5.

Check It Out

Find the median.

5 21, 38, 15, 8, 18, 21, 8

6 24, 26, 2, 33

7 90, 96, 68, 184, 176, 86, 116

8 Yeaphana measured the weight of 10 adults, in pounds: 160, 140, 175, 141, 138, 155, 221, 170, 150, and 188. Find the median weight.

Mode

You can also describe the central tendency of a set of numbers by using the mode. The **mode** is the number in the set that occurs most frequently. Let's look again at the allowances:

$1 $1 $2 $3 $3 $3 $3 $5 $6 $6 $7 $7 $8 $8 $42

To find the mode, look for the number that appears most frequently.

EXAMPLE   Finding the Mode

Find the mode of the allowances.

- Arrange the numbers in order or make a frequency table of the numbers.

    The numbers are arranged in order above.

- Select the number that appears most frequently.

    The most common allowance is $3.

So, Mei Mei was using the mode when she described the typical allowance.

A group of numbers may have no mode or more than one mode. Data that have two modes is called *bimodal*.

Check It Out

Find the mode.

9 53, 52, 56, 53, 53, 52, 57, 56

10 100, 98, 78, 98, 96, 87, 96

11 12, 14, 14, 16, 21, 15, 14, 13, 20

12 Attendance at the zoo one week was as follows: 34,543; 36,122; 35,032; 36,032; 23,944; 45,023; 50,012.

Outliers

Values that are much higher or lower than others in a data set are called *outliers*. You can see how an outlier affects measures of central tendency by calculating those measures with and without the outlier. The data set for the allowances has one outlier that is much higher than the rest of the values in the set.

$1, $1, $2, $3, $3, $3, $3, $5, $6, $6, $7, $7, $8, $8, **$42**

EXAMPLE **Determining How Outliers Affect Central Tendency**

Compare the measures of central tendency for the allowances with and without the outlying value of $42. For this data set, including the outlier: the mean is $7; the median is $5; and the mode is $3.

Now find the values of these measures *without the outlier.*

- To find the mean, add the amounts shown above, and divide the total by the number of allowances.

    $63 ÷ 14 = 4.5

The new mean allowance is $4.50.

- To find the median, look at the allowances in numerical order from least to greatest. Since there are an even number of values in this set, find the mean of the two middle numbers.

    (3 + 5) ÷ 2 = 4

The new median is $4.00.

- To find the mode, look at the numbers arranged in order (as above). Select the number that appears most frequently.

The most common allowance is $3.
The mode remains $3.00.

The outlying allowance of $42 does not affect the mode. It has a small affect on the median. The outlier has the greatest affect on the mean of a set of numbers. The mean calculated without the outlier better represents the data.

Find and compare means.

13 Find the mean of this set of data. 23, 2, 21, 23, 19, 18, 20

14 Identify the outlier in the data set in Exercise 13.

15 Find the mean of the data set in Exercise 13 without the outlier. Describe how the outlier affects the mean.

Range

Another measure used with numbers is the range. The **range** is the difference between the greatest and least number in a set. Consider the seven tallest buildings in Phoenix, Arizona:

Building Height	
Building 1	407 ft
Building 2	483 ft
Building 3	372 ft
Building 4	356 ft
Building 5	361 ft
Building 6	397 ft
Building 7	397 ft

To find the range, you must subtract the shortest height from the tallest.

EXAMPLE **Find the Range**

Find the range of the building heights in Phoenix.

- Find the greatest and least values.

    The greatest value is 483 and the least value is 356.

- Subtract.

    $483 - 356 = 127$

The range is 127 feet.

Check It Out

Find the range.

16 110, 200, 625, 300, 12, 590

17 24, 35, 76, 99

18 23°, 6°, 0°, 14°, 25°, 32°

APPLICATION — How Mighty Is the Mississippi?

The legendary Mississippi is the longest river in the United States, but not in the world. Here's how it compares to some of the world's longest rivers.

Mississippi

River	Location	Length (miles)
Nile	Africa	4,160
Amazon	South America	4,000
Yangtze	Asia	3,964
Yellow	Asia	3,395
Ob-Irtysh	Asia	3,362
Congo	Africa	2,718
Mekong	Asia	2,600
Niger	Africa	2,590
Yenisey	Asia	2,543
Parana	South America	2,485
Mississippi	North America	2,340
Missouri	North America	2,315

In this set of data, what is the mean length, the median length, and the range? See **Hot**Solutions for answer.

4·3 Exercises

Find the mean, median, mode, and range. Round to the nearest tenth.

1. 2, 4, 5, 5, 6, 6, 7, 7, 7, 9
2. 18, 18, 20, 28, 20, 18, 18
3. 14, 13, 14, 15, 16, 17, 23, 14, 16, 20
4. 79, 94, 93, 93, 80, 86, 82, 77, 88, 90, 89, 93
5. Are any of the sets of data in Exercises 1–4 bimodal? Explain.

6. When a cold front went through Lewisville, the temperature dropped from 84° to 38°. What was the range in temperatures?

7. In one week, there were the following number of accidents in Caswell: 1, 1, 3, 2, 5, 2, and 1. Which of the measures (mean, median, or mode) do you think you should use to describe the number of accidents? Explain.

8. Does the median have to be a member of the set of data?

9. Are you using the mean, median, or mode when you say that half the houses in Sydneyville cost more than $150,000?

10. The following numbers represent the numbers of phone calls received each hour between noon and midnight during one day at a mail-order company.

    13 23 14 12 80 22 14 25 14 17 12 18

    Find the mean, median, and mode of the calls. Which measure best represents the data? Explain.

11. What is the outlier in the numbers of calls shown in Exercise 10?

12. What is the difference in means for the numbers of calls in Exercise 10 calculated with and without the outlier?

4·4 Probability

If you and a friend want to decide who goes first in a game, you might flip a coin. You and your friend each have an equal chance of winning the toss. The **probability** of an event is a number between 0 and 1 that measures the chance that an event will occur. The closer a probability is to 1, the more likely it is that the event will happen. You can use a ratio to express a probability.

Simple Events

The probability of an event is the ratio that compares the number of favorable outcomes to the number of possible outcomes. A **simple event** is a specific outcome or type of outcome.

EXAMPLE **Determining the Probability of a Simple Event**

You and a friend decide who goes first by flipping a coin. What is the probability that you will get to go first?

$$\text{probability of an event} = \frac{\text{number of favorable outcomes}}{\text{number of possible outcomes}}$$

- In a coin toss, there are two possible outcomes—heads or tails.
- Suppose you call "heads"; there is one favorable outcome.

Possible Outcomes	Event or Favorable Outcome
2	1

The probability that you win the coin toss is $\frac{1}{2}$.

Express each of the following probabilities as a fraction.

1 the probability of rolling an even number on a 6-sided die with sides numbered 1 through 6

2 the probability of rolling a number greater than 4 on a 12-sided die with sides numbered 1 through 12

Expressing Probabilities

You can express a probability as a fraction, as shown before. But just as you can write a fraction as a decimal, ratio, or percent, you can also write a probability in any of these forms (p. 147).

The probability of showing an even number after you roll a 1–6 number cube is $\frac{1}{2}$. You can also express the probability as follows:

Fraction	Decimal	Ratio	Percent
$\frac{1}{2}$	0.5	1:2	50%

Check It Out

Express each of the following probabilities as a fraction, decimal, ratio, and percent.

3 the probability of drawing a spade when drawing a card from a deck of cards

4 the probability of getting a green counter when drawing a counter from a bag containing 4 green counters and 6 white ones

Probability Line

You know that the probability of an event is a number between 0 and 1. One way to show how probabilities relate to each other is to use a **probability line**. The following probability line shows the possible ranges of probability values:

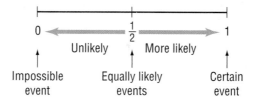

The line shows that events that are certain to happen have a probability of 1. Such an event is the probability of rolling a 1, 2, 3, 4, 5, or 6 when rolling a standard number cube. An event that cannot happen has a probability of 0. The probability of getting an 8 when spinning a spinner that shows 0, 2, and 4 is 0. Events that are equally likely, such as getting a head or tail when you toss a coin, have a probability of $\frac{1}{2}$.

Use a number line to compare the probability of randomly drawing a black marble with the probability of randomly drawing a blue marble from a bag of 8 white marbles and 12 black marbles.

EXAMPLE Comparing Probabilities on a Probability Line

- Draw a number line and label it from 0 to 1.
- Calculate the probabilities of the given events and show them on the probability line.

  The probability of getting a black marble is $\frac{12}{20} = \frac{3}{5}$ and the probability of getting a blue marble is zero. The probabilities are shown on the following probability line.

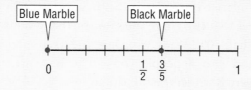

Draw a probability line. Then plot the following:

⑤ the probability of drawing a white token from a bag of white tokens

⑥ the probability of tossing two heads on two tosses of a coin

4·4

PROBABILITY

APPLICATION **Birthday Surprise**

How likely do you think it is that two people in your class have the same birthday? With 365 days in a year, you might think the chances are very slim. After all, the probability that a person is born on any given day is $\frac{1}{365}$, or about 0.3%.

Try taking a survey. Ask your classmates to write their birthdays on separate slips of paper. Don't forget to write your birthday, too. Collect the slips and see if any birthdays match.

It might surprise you to learn that in a group of 23 people, the chances that two share the same birthday is just a slight bit more than 50%. With 30 people, the likelihood increases to 71%. And with 50 people, you can be 97% sure that two of them were born on the same date.

Sample Spaces

The set of all possible outcomes is called the **sample space**. You can make a list, diagram, or grid to determine the sample space.

Tree Diagrams

You often need to be able to count outcomes. For example, suppose that you have a **spinner** that is half red and half green. You can make a **tree diagram** to find all the possible outcomes if you spin the spinner three times.

EXAMPLE Making a Tree Diagram

Determine the sample space of spinning the red and green spinner three times.

- List the possible outcomes of the first, second, and third spins.

  The spinner can show red or green.

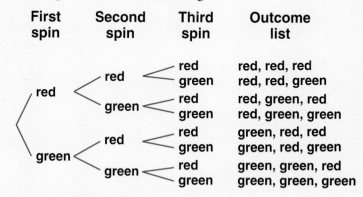

First spin	Second spin	Third spin	Outcome list
red	red	red	red, red, red
	red	green	red, red, green
	green	red	red, green, red
	green	green	red, green, green
green	red	red	green, red, red
	red	green	green, red, green
	green	red	green, green, red
	green	green	green, green, green

- Draw lines and list all the possible outcomes.

  The outcomes are listed above. There are eight different outcomes.

You can also find the total number of possibilities by multiplying the number of choices at each step: $2 \times 2 \times 2 = 8$.

Check It Out

Draw a tree diagram for Exercises 7–10. Check by multiplying.

7 Kimiko can choose a sugar cone, a regular cone, or a dish. She can have vanilla, chocolate, or strawberry yogurt. How many possible desserts can she have?

8 Kirti sells round, square, oblong, and flat beads. They come in green, orange, and white. If he separates all the beads by color and shape, how many containers does he need?

9 If you toss three coins at once, how many possible ways can they land?

10 How many possible ways can Norma climb to Mt. Walker if she goes through Soda Spring?

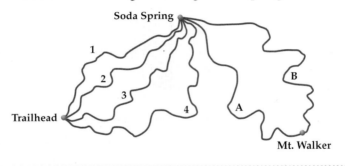

Outcome Grids

You have seen how to use a tree diagram to show possible outcomes. Another way to show the outcomes in an experiment is to use an **outcome grid**. The following outcome grid shows the outcomes when tossing a coin two times:

	Head	Tail
Head	head, head	head, tail
Tail	tail, head	tail, tail

You can use the grid to find the number of ways the coins can come up the same, which is two.

EXAMPLE Making Outcome Grids

Make an outcome grid to show the possible outcome of rolling two standard number cubes and then adding the two numbers together.

- List the outcomes of the first type down the side. List the outcomes of the second type across the top.
- Fill in the outcomes.

	1	2	3	4	5	6
1	2	3	4	5	6	7
2	3	4	5	6	7	8
3	4	5	6	7	8	9
4	5	6	7	8	9	10
5	6	7	8	9	10	11
6	7	8	9	10	11	12

Once you have completed the outcome grid, it is easy to count target outcomes and determine probabilities. So, what is the probability that you will roll a sum of 8?

- Count the possible outcomes.

    There are 36 outcomes listed in the chart.

- Count how many times the sum of 8 is in the chart.

    8 is listed 5 times.

The probablity of rolling the sum of 8 is $\frac{5}{36}$.

Use the spinner drawing to answer Exercises 11 and 12.

11 Make an outcome grid to show the outcomes when making a two-digit number by spinning the spinner twice.

12 What is the probability of getting a number divisible by two when you spin the spinner twice?

Experimental Probability

The theoretical probability of an event is a number between 0 and 1. One way to find the probability of an event is to conduct an experiment. Suppose that you want to know the probability of seeing a friend when you ride your bike. You ride your bike 20 times and see a friend 12 times. You can compare the number of times you see a friend to the number of times you ride your bike to find the probability of seeing a friend. In this case, the **experimental probability** that you will see a friend is $\frac{12}{20}$, or $\frac{3}{5}$.

EXAMPLE Determining Experimental Probability

Find the experimental probability of getting heads more than tails in 100 coin tosses.

- Conduct an experiment. Record the number of trials and the result of each trial.

  Toss the coin 100 times and count the number of heads and tails. Suppose that you get heads 36 times and tails 62 times.

- Compare the number of occurrences of one result to the number of trials. That is the probability for that result.

The experimental probability of getting heads in 100 coin tosses is $\frac{36}{100}$, or $\frac{9}{25}$.

Slips of paper labeled red, green, yellow, and blue are
drawn from a bag containing 40 slips of paper. Use
the results shown on the circle graph to answer Exercises
13 and 14.

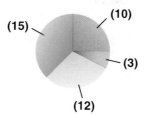

13 Find the experimental probability of getting blue.

14 Find the experimental probability of getting either
yellow or red.

15 Toss a coin 50 times. Find the experimental probability
of getting a head. Compare your answers with others'
answers.

Theoretical Probability

You know that you can find the experimental probability of
tossing a head when you toss a coin by doing the experiment
and recording the results. You can also find the **theoretical
probability** by considering the outcomes of the experiment.
The **outcome** of an experiment is a result. The possible outcomes
when rolling a standard number cube are the numbers 1–6.
An **event** is a specific outcome, such as 5. So the probability
of getting a 5 is:

$$\frac{\text{number of ways an event occurs}}{\text{number of possible outcomes}} = \frac{1 \text{ way to get a } 5}{6 \text{ possible outcomes}} = \frac{1}{6}$$

EXAMPLE	Determining Theoretical Probability

Find the probability of rolling a two or a three when you roll a number cube labeled 1–6.

- Determine the total number of ways the event occurs.

   In this case, the event is getting a two or a three. There are two ways to get a two or three.

- Determine the total number of outcomes. Use a list, multiply, or make a *tree diagram* (p. 199).

   There are six numbers on the cube.

- Use the formula.

   $$P(\text{event}) = \frac{\text{number of ways an event occurs}}{\text{number of outcomes}}$$

   Find the probability of rolling a two or a three, represented by $P(2 \text{ or } 3)$, by substituting into the formula.

   $$P(2 \text{ or } 3) = \frac{2}{6} = \frac{1}{3}$$

   The probability of rolling a two or a three is $\frac{1}{3}$.

Check It Out

Find each probability. Use the spinner drawing for Exercises 16 and 17.

16 $P(3)$

17 $P(1, 2, 3, \text{ or } 4)$

18 $P(\text{even number})$ when rolling a 1–6 number cube

19 The letters of the words *United States* are written on slips of paper, one to a slip, and placed in a bag. If you draw a slip at random, what is the probability it will be a vowel?

4·4 Exercises

You spin a spinner divided into eight equal parts numbered 1–8. Find each probability as a fraction, a decimal, a ratio, and a percent.

1. P(odd number)

2. P(1 or 2)

3. If you pick a marble from a bag of marbles containing 4 red and 6 black marbles without looking, what is the probability of picking a red marble? Is this experimental or theoretical probability?

4. If you toss a thumbtack 50 times and it lands up 15 times, what is the probability of it landing up? Is this experimental or theoretical probability?

5. Draw a probability line to show the probability of getting a number less than 7 when rolling a standard number cube.

6. If you have a spinner with red, blue, green, and yellow sections of equal size, how many different possible outcomes can you get as a result of four spins?

7. Make an outcome grid to show the outcomes of spinning a spinner containing the numbers 1–4 and tossing a coin.

Data, Statistics, and Probability

What have you learned?

You can use the problems and the list of words that follow to see what you learned in this chapter. You can find out more about a particular problem or word by referring to the topic number (*for example*, Lesson 4·2).

Problem Set

1. Laila listed the businesses in her community on slips of paper and then drew 20 names to choose businesses to survey. Is this a random sample? (Lesson 4·1)

2. A survey asked, "Do you care enough about the environment to recycle?" Is this question biased? If so, rewrite it. (Lesson 4·1)

3. What type of display organizes data into columns and rows to show frequency? (Lesson 4·1)

4. Is the following statement true or false? Explain. A circle graph shows how part of a group is divided. (Lesson 4·2)

Use the following line graph, which shows the average low temperatures each month in Seattle, to answer Exercises 5 and 6. (Lesson 4·2)

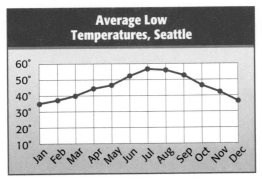

5. During which months is the average low less than 40°?

6. What is the range of low temperatures?

7. A display of X's above a number line to represent tally marks is called what? (Lesson 4·2)

8. Is the following statement true or false? Explain. A bar graph shows how a whole is divided. (Lesson 4·2)

9. Find the mean, median, mode, and range of the numbers 45, 35, 43, 26, and 21. (Lesson 4·3)

10. Must the median be a member of the set of data? (Lesson 4·3)

11. What are the mean, median, and mode measures of? (Lesson 4·3)

12. What is the term for a member of a set of data that is significantly greater or less than the rest of the set? (Lesson 4·3)

13. What is the probability of rolling a sum of seven when you roll two number cubes? (Lesson 4·4)

HotWords
Write definitions for the following words.

average (Lesson 4·3)
bar graph (Lesson 4·2)
circle graph (Lesson 4·2)
experimental probability (Lesson 4·4)
leaf (Lesson 4·2)
line graph (Lesson 4·2)
mean (Lesson 4·3)
median (Lesson 4·3)
mode (Lesson 4·3)
outcome (Lesson 4·4)
outcome grid (Lesson 4·4)
outlier (Lesson 4·3)
percent (Lesson 4·2)
population (Lesson 4·1)

probability (Lesson 4·4)
probability line (Lesson 4·4)
random sample (Lesson 4·1)
range (Lesson 4·3)
sample (Lesson 4·1)
simple event (Lesson 4·4)
spinner (Lesson 4·4)
stem (Lesson 4·2)
stem-and-leaf plot (Lesson 4·2)
survey (Lesson 4·1)
table (Lesson 4·1)
tally marks (Lesson 4·1)
theoretical probability (Lesson 4·4)
tree diagram (Lesson 4·4)

HotTopic 5

Algebra

What do you know? You can use the problems and the list of words that follow to see what you already know about this chapter. The answers to the problems are in **HotSolutions** at the back of the book, and the definitions of the words are in **HotWords** at the front of the book. You can find out more about a particular problem or word by referring to the topic number (*for example,* Lesson 5·2).

Problem Set

Write an expression for each phrase. (Lesson 5·1)

1. a number increased by five

2. the product of four and some number

3. two less than the sum of a number and three

Simplify each expression. (Lesson 5·2)

4. $8x + 7 - 5x$

5. $4(2n - 1) - (n + 6)$

6. Evaluate the expression $6x + 12$ for $x = 4$. (Lesson 5·3)

7. Write an equation for the following: three times a number added to five is the same as 22. (Lesson 5·4)

8. Determine whether $11 + p = 16$ is true or false for $p = 3$, $p = 4$, and $p = 5$. (Lesson 5·4)

9. Find the distance traveled by a jogger who jogs at 6 miles per hour for $2\frac{1}{2}$ hours. (Lesson 5·5)

Use a proportion to solve the problem. (Lesson 5·5)

10. In a class, the ratio of boys to girls is $\frac{2}{3}$. If there are 15 girls in the class, how many boys are there?

11. A map is drawn using a scale of 80 kilometers to 1 centimeter. The distance between two cities is 600 kilometers. How far apart are the two cities on the map?

Solve each inequality. (Lesson 5·6)

12. $x + 4 < 7$ **13.** $3x \geq 12$ **14.** $n - 8 > -6$ **15.** $\frac{n}{2} \leq -1$

Locate each point on a coordinate plane and tell in which quadrant or on which axis it lies. (Lesson 5·7)

16. $A(1, 3)$ **17.** $B(-3, 0)$ **18.** $C(3, -2)$

19. $D(0, -4)$ **20.** $E(-2, 5)$ **21.** $F(-4, -3)$

22. Make a function table for the rule $y = x + 4$ using input values 1, 2, and 3. Plot the points on a coordinate plane. (Lesson 5·7)

algebra (Lesson 5·1)

Associative Property
(Lesson 5·2)

axes (Lesson 5·7)

Commutative Property
(Lesson 5·2)

cross product (Lesson 5·5)

difference (Lesson 5·1)

Distributive Property
(Lesson 5·2)

equation (Lesson 5·4)

equivalent (Lesson 5·4)

equivalent expression
(Lesson 5·2)

expression (Lesson 5·1)

formula (Lesson 5·3)

function (Lesson 5·4)

horizontal (Lesson 5·7)

inequality (Lesson 5·6)

like terms (Lesson 5·2)

order of operations (Lesson 5·3)

ordered pair (Lesson 5·7)

origin (Lesson 5·7)

perimeter (Lesson 5·3)

point (Lesson 5·7)

product (Lesson 5·1)

proportion (Lesson 5·5)

quadrant (Lesson 5·7)

quotient (Lesson 5·1)

rate (Lesson 5·5)

ratio (Lesson 5·5)

solution (Lesson 5·4)

sum (Lesson 5·1)

term (Lesson 5·1)

unit rate (Lesson 5·5)

variable (Lesson 5·1)

vertical (Lesson 5·7)

x-axis (Lesson 5·7)

y-axis (Lesson 5·7)

5·1 Writing Expressions and Equations

Expressions

A **variable** is a symbol, usually a letter, that is used to represent an unknown number. Here are some commonly used variables:

$$x \quad n \quad y \quad a \quad ?$$

Algebra is a language of symbols, including variables. An algebraic **expression** is a combination of numbers, variables, and at least one operation (addition, subtraction, multiplication, or division).

$4 + x$ means 4 plus *some number.*

$4x$ means 4 times *some number.*

Each number, variable, or number and variable together in an expression is called a **term**.

The terms in the expression $4 + x$ are 4 and x.

The combination $4x$ is a single term.

Check It Out

Count the number of terms in each expression.

1 $6n + 4$

2 $7bc$

3 $5m - 3n - 4$

4 $\frac{x}{4}$

Writing Expressions Involving Addition

To write an expression, you often have to interpret a written phrase. For example, the phrase "four added to some number" can be written as the expression $x + 4$, where the variable x represents the unknown number.

The words "added to" indicate that the operation between 4 and the number is addition. Other words and phrases that indicate addition are "more than," "plus," and "increased by." Another word that indicates addition is **sum**. The sum is the result of addition.

Some common addition phrases and the corresponding expressions are listed below.

Phrase	Expression
3 more than some number	$n + 3$
a number increased by 7	$x + 7$
9 plus some number	$9 + y$
the sum of a number and 6	$n + 6$

➡ **Check It Out**

Write an expression for each phrase.

5 a number added to 5

6 the sum of a number and 4

7 some number increased by 8

8 2 more than some number

Writing Expressions Involving Subtraction

The phrase "four subtracted from some number" can be written as the expression $x - 4$. The words "subtracted from" indicate that the operation between the unknown number and four is subtraction.

Some other words and phrases that indicate subtraction are "less than," "minus," and "decreased by." Another word that indicates subtraction is **difference**. The difference between two terms is the result of subtraction.

In a subtraction expression, the order of the terms is very important. You have to know which term is being subtracted and which is being subtracted from. To help interpret the phrase "six less than a number," replace "a number" with 10. What is 6 less than 10? The answer is 4, which is $10 - 6$, not $6 - 10$. The phrase translates to the expression $x - 6$, not $6 - x$.

Some common subtraction phrases and the corresponding expressions are listed below.

Phrase	Expression
5 less than some number	$n - 5$
a number decreased by 8	$x - 8$
7 minus some number	$7 - n$
the difference between a number and 2	$n - 2$

⇒ **Check It Out**

Write an expression for each phrase.

9 a number subtracted from 8

10 the difference between a number and 3

11 some number decreased by 6

12 4 less than some number

Writing Expressions Involving Multiplication

The phrase "four multiplied by some number" can be written as the expression $4x$. The words "multiplied by" indicate that the operation between the unknown number and four is multiplication.

Some other words and phrases that indicate multiplication are "times," "twice," and "of." "Twice" is used to mean "two times." "Of" is used primarily with fractions and percents. Another word that indicates multiplication is **product**. The product of two terms is the result of multiplication.

The operation of multiplication can be represented in several different ways. All of the following are ways to write "5 multiplied by some number":

$5 \times n$

$5n$

$5(n)$

$5 \cdot n$

Some common multiplication phrases and the corresponding expressions are listed below.

Phrase	Expression
5 times some number	$5a$
twice a number	$2x$
one-fourth of some number	$\frac{1}{4}y$
the product of a number and 8	$8n$

Check It Out

Write an expression for each phrase.

13 a number multiplied by 4

14 the product of a number and 8

15 25% of some number

16 9 times some number

Multiplication is commutative; therefore, the expressions $4x$ and $x \cdot 4$ are equivalent. However, when writing an expression that involves multiplying a number and a variable, it is considered standard form for the number to precede the variable.

Writing Expressions Involving Division

The phrase "four divided by some number" can be written as the expression $\frac{4}{x}$. The words "divided by" indicate that the operation between the unknown number and four is division.

Some other words and phrases that indicate division are "ratio of" and "divide." Another word that indicates division is **quotient**. The quotient of two terms is the result of division.

Like subtraction, the order of the terms in a division expression is very important. When writing a division expression as a fraction, the number being divided is always the numerator, and the number being divided by is always the denominator. The phrase "12 divided by 3" is written $\frac{12}{3}$. Switching the numerator and denominator will result in a different answer.

$$\frac{12}{3} = 4 \text{ and } \frac{3}{12} = 0.25$$

$$\text{Therefore, } \frac{12}{3} \neq \frac{3}{12}.$$

Some common division phrases and the corresponding expressions are listed below.

Phrase	Expression
the quotient of 20 and some number	$\frac{20}{n}$
a number divided by 6	$\frac{x}{6}$
the ratio of 10 and some number	$\frac{10}{y}$
the quotient of a number and 5	$\frac{n}{5}$

➡️ **Check It Out**

Write an expression for each phrase.

17 a number divided by 5

18 the quotient of 8 and a number

19 the ratio of 20 and some number

20 the quotient of some number and 4

5·1 Exercises

Count the number of terms in each expression.

1. $8x + n$

2. $5x$

3. $6x - 3y + 5z$

4. $37 + 2x$

Write an expression for each phrase.

5. two more than a number

6. a number added to six

7. the sum of a number and four

8. five less than a number

9. 12 decreased by some number

10. the difference between a number and three

11. one third of some number

12. twice a number

13. the product of a number and eight

14. a number divided by seven

15. the ratio of 10 and some number

16. the quotient of a number and six

17. Which of the following words is used to indicate multiplication?
    A. sum
    B. difference
    C. product
    D. quotient

18. Which of the following does not indicate subtraction?
    A. less than
    B. difference
    C. decreased by
    D. ratio of

19. Which of the following shows "twice the sum of a number and four"?
    A. $2(x + 4)$
    B. $2x + 4$
    C. $2(x - 4)$
    D. $2 + (x + 4)$

20. In which of the following operations is the order of the terms important?
    A. addition
    B. division
    C. multiplication
    D. all of them

5·2 Simplifying Expressions

Terms

Terms are numbers, variables, or numbers and variables combined by multiplication or division. The following are some examples of terms:

$$n \qquad 7 \qquad 5x \qquad x^2 \qquad 3(n + 5)$$

Compare the terms 7 and $5x$. The value of $5x$ will change as the value of x changes. If $x = 2$, then $5x = 5(2) = 10$; and if $x = 3$, then $5x = 5(3) = 15$. Notice, though, that the value of 7 never changes—it remains constant. When a term contains a number only, it is called a *constant* term.

➡ **Check It Out**

Decide whether each term is a constant term.

1 $6x$

2 9

3 $3(n + 1)$

4 5

The Commutative Property of Addition and Multiplication

The **Commutative Property** of Addition states that the order of terms being added may be switched without changing the result: $3 + 4 = 4 + 3$, and $x + 8 = 8 + x$. The Commutative Property of Multiplication states that the order of terms being multiplied may be switched without changing the result; $3(4) = 4(3)$ and $x \cdot 8 = 8x$.

The Commutative Property does not hold for subtraction or division. The order of the terms does affect the result: $5 - 3 = 2$, but $3 - 5 = -2$; $8 \div 4 = 2$, but $4 \div 8 = \frac{1}{2}$.

Rewrite each expression, using the Commutative Property of Addition or Multiplication.

5 $2x + 7$ **6** $n \cdot 6$ **7** $5 + 4y$ **8** $3 \cdot 8$

The Associative Property of Addition and Multiplication

The **Associative Property** of Addition states that the grouping of terms being added does not affect the result:
$(3 + 4 + 5 = 3 + (4 + 5)$, and $(x + 6) + 10 = x + (6 + 10)$.

The Associative Property of Multiplication states that the grouping of terms being multiplied does not affect the result:
$(2 \cdot 3) \cdot 4 = 2 \cdot (3 \cdot 4)$, and $5 \cdot 3x = (5 \cdot 3)x$.

The Associative Property does not hold for subtraction or division. The grouping of the numbers does affect the result:
$(8 - 6) - 4 = -2$, but $8 - (6 - 4) = 6$.
$(16 \div 8) \div 2 = 1$, but $16 \div (8 \div 2) = 4$.

APPLICATION ... 3, 2, 1, Blast Off

Fleas, those tiny pests that make your dog or cat itch, are amazing jumpers. Actually, fleas don't jump but launch themselves at a rate 50 times faster than the space shuttle leaves Earth.

Elastic pads on the flea's feet compress like a coiled spring to power a liftoff. When a flea is ready to leave, it hooks onto its host, locks its legs in place, then releases the "hooks." The pads spring back into shape and the flea blasts off. A flea no bigger than 0.05 of an inch can propel itself a distance of as much as 8 inches—160 times its own length! If you could match a flea's feat, how far could you jump? See **HotSolutions** for the answer.

➡️ **Check It Out**

Rewrite each expression, using the Associative Property of Addition or Multiplication.

9 $(4 + 5) + 8$

10 $(2 \cdot 3) \cdot 5$

11 $(5x + 4y) + 3$

12 $6 \cdot 9n$

The Distributive Property

The **Distributive Property** of Addition and Multiplication states that multiplying a sum by a number is the same as multiplying each addend by that number and then adding the two products.

$$3 \times (2 + 3) = (3 \times 2) + (3 \times 3)$$

How would you multiply 7×99 in your head? You might think $700 - 7 = 693$. If you did, you have used the Distributive Property.

$7(100 - 1)$

- Distribute the factor of 7 to each term inside the parentheses.

$= (7 \cdot 100) - (7 - 1)$

- Simplify, using order of operations.

$= 700 - 7$

$= 693$

The Distributive Property does not hold for division.

$3 \div (2 + 3) \neq (3 \div 2) + (3 \div 3)$.

➡️ **Check It Out**

Use the Distributive Property to find each product.

13 $6 \cdot 99$

14 $3 \cdot 106$

15 $4 \cdot 198$

16 $5 \cdot 211$

Equivalent Expressions

The Distributive Property can be used to write an **equivalent expression**. Equivalent expressions are different ways of writing one expression.

EXAMPLE   Writing an Equivalent Expression

Write an equivalent expression for $5(9x - 7)$.

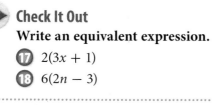

$5 \times 9x - 5 \times 7$	• Distribute the factor of 5 to each term inside the parentheses.
$45x - 35$	• Simplify.
$5(9x - 7) = 45x - 35$	• Write the equivalent expressions.

Check It Out

Write an equivalent expression.

17 $2(3x + 1)$

18 $6(2n - 3)$

Like Terms

Like terms are terms that contain the same variables. Constant terms are like terms because they do not have any variables. Two examples of like terms are listed below.

Like Terms	Reason
$3x$ and $4x$	Both contain the same variables.
3 and 11	Both are constant terms.

Two examples of terms that are not like terms are listed below.

Not Like Terms	Reason
$3x$ and $5y$	Variables are different.
$4n$ and 12	One term has a variable; the other is constant.

Terms that have more than one variable can also be like terms. Each term must contain an identical set of variable factors.

Like Terms	Reason
xyz and *xyz*	Both contain the same variable factors.
ab and 2*ab*	The numeric factor is different, but the variable factors are identical.
7*cd* and 3*dc*	Multiplication is commutative, so 3*dc* can be written 3*cd*. The variable factors are identical.

Two examples of terms that are not like terms are listed below.

Not Like Terms	Reason
4*rst* and 4*rt*	The variable factors are not the same.
wx and 9	One term contains variables; the other is constant.

Like terms may be combined into one term by adding or subtracting.

EXAMPLE Combining Like Terms

Simplify $5n - 3n + 4n$.

$2n + 4n$ • Use order of operations. Subtract like terms.

$6n$ • Add like terms.

So, the like terms $5n - 3n + 4n$ can be simplified to $6n$.

Check It Out
Combine like terms.

19. $5x + 6x$
20. $10y - 4y$
21. $4n + 3n + n$
22. $2a + 8a$

Simplifying Expressions

Expressions are simplified when all of the like terms have been combined. Terms that are not like terms cannot be combined.

In the expression $6x + 5y - 3x$, there are three terms. Two of them are like terms, $6x$ and $3x$. Rewrite the expression, using the Commutative Property: $5y + 6x - 3x$. Subtract the like terms: $5y + 3x$. The expression is simplified because the two remaining terms are not like terms.

EXAMPLE Simplifying Expressions

Simplify the expression $(2 + z) + 7$.

$(2 + z) + 7$

$= (z + 2) + 7$ • Use the Commutative Property.

$= z + (2 + 7)$ • Use the Associative Property.

$= z + 9$ • Add 2 and 7.

So, $(2 + z) + 7$ can be simplified to $z + 9$.

Simplify the expression $9(10p)$.

$9(10p)$

$= 9 \cdot (10 \cdot p)$ • Parentheses mean to multiply.

$= (9 \cdot 10) \cdot p$ • Use the Associative Property.

$= 90p$ • Multiply 9 and 10.

So, $9(10p)$ can be simplified to $90p$.

Check It Out

Simplify each expression.

23 $2 + (6 + y)$

24 $(15 + n) + 22$

25 $5(4x)$

26 $(3 \times n) \times 12$

5·2 Exercises

Decide whether each term is a constant term.

1. $6n$ 2. -5

Using the Commutative Property of Addition or Multiplication, rewrite each expression.

3. $3 + 8$ 4. $n \cdot 5$ 5. $4x + 7$

Using the Associative Property of Addition or Multiplication, rewrite each expression.

6. $3 + (6 + 8)$ 7. $(4 \cdot 5) \cdot 3$

8. $5 \cdot 2n$

Use the Distributive Property to find each product.

9. $8 \cdot 99$ 10. $3 \cdot 106$

Write an equivalent expression.

11. $3(4x + 1)$ 12. $5(2n + 3)$

13. $10(3b - 4)$ 14. $3(5y - 2)$

Combine like terms.

15. $10x - 3x$ 16. $4m - 6m$

17. $5n + 4n - n$

Simplify each expression.

18. $5n + 3b - n$ 19. $2x + 2(3x)$

20. $2(3n - 1)$

21. Which property is illustrated by $5(2x + 1) = 10x + 5$?
    A. Commutative Property of Multiplication
    B. Distributive Property
    C. Associative Property of Multiplication
    D. The example does not illustrate a property.

22. Which of the following is an equivalent expression to $24x - 36$?
    A. $2(12x + 18)$ B. $3(8x + 33)$
    C. $6(4x - 6)$ D. $24(x - 12)$

5·3 Evaluating Expressions and Formulas

Evaluating Expressions

Once an expression has been written, you can evaluate it for different values of the variable. Remember—always use **order of operations** to evaluate expressions. Multiply first, and then subtract.

To evaluate $5x - 3$ for $x = 4$, *substitute* 4 in place of the x: $5(4) - 3$. So $5(4) - 3 = 20 - 3 = 17$.

EXAMPLE Evaluating Expressions

Evaluate $3(x - 2) + 5$ for $x = 4$.

$3(4 - 2) + 5$	• Substitute the value for x.
$3(2) + 5$	• Use order of operations to simplify. Simplify within parentheses first.
$6 + 5$	• Multiply.
11	• Add.

So, when $x = 4$, $3(x - 2) + 5 = 11$.

Check It Out

Evaluate each expression for the given value.

1. $4x - 5$ for $x = 4$
2. $2a + 5$ for $a = 3$
3. $6(n - 5) + 3$ for $n = 9$
4. $2(3y - 2)$ for $y = 2$

Evaluating Formulas

The Formula for Perimeter of a Rectangle

The **perimeter** of a rectangle is the distance around the rectangle. The formula $P = 2w + 2\ell$ can be used to find the perimeter, P, if the width, w, and the length, ℓ, are known.

EXAMPLE Finding the Perimeter of a Rectangle

Find the perimeter of a rectangle whose width is 3 feet and length is 4 feet.

$P = 2w + 2\ell$ • Use the formula for the perimeter of a rectangle ($P = 2w + 2\ell$).

$= (2 \times 3) + (2 \times 4)$ • Substitute the given values for w and ℓ.

$= 6 + 8$ • Multiply.

$= 14$ • Add.

The perimeter of the rectangle is 14 feet.

The formula for the perimeter of a rectangle can be simplified using the distributive property.

$$P = 2w + 2\ell$$
$$= 2(w + \ell)$$

EXAMPLE Finding Perimeter Using the Formula $P = 2(w + \ell)$

Find the perimeter of a rectangle whose width is 7 meters and length is 3 meters, using $P = 2(w + \ell)$.

$P = 2(w + \ell)$ • Use the simplified formula for the perimeter of a rectangle $P = 2(w + \ell)$.

$P = 2(7 + 3)$ • Substitute the given values for w and ℓ.

$= 2(10)$ • Use order of operations to simplify. Simplify within the parentheses first.

$= 20$ • Multiply.

The perimeter of the rectangle is 20 meters.

Check It Out

Find the perimeter of each rectangle described.

5 $w = 6$ cm, $\ell = 10$ cm

6 $w = 2.5$ ft, $\ell = 6.5$ ft

The Formula for Distance Traveled

The distance traveled by a person, vehicle, or object depends on its rate of travel and the amount of time it travels. The formula $d = rt$ can be used to find the distance traveled, d, if the rate, r, and the amount of time, t, are known.

EXAMPLE Finding the Distance Traveled

Find the distance traveled by a bicyclist who averages 20 miles per hour for 4 hours.

$d = 20 \cdot 4$ • Substitute values in the distance formula $(d = rt)$.

$d = 80$ • Multiply.

The bicyclist rode 80 miles.

The cheetah is the fastest land animal, reaching speeds up to 70 miles per hour. That is an average speed of about 103 feet per second. Cheetahs can maintain this rate for about 9 seconds. About how far can a cheetah run in 9 seconds?

$d = rt$

$d = 103 \cdot 9$ • Substitute the values in the formula.

$d = 927$ • Multiply.

Cheetahs can run about 927 feet in 9 seconds.

Check It Out

Find the distance traveled, if

7 a person rides 10 miles per hour for 3 hours.

8 a plane flies 600 kilometers per hour for 2 hours.

9 a person drives a car 55 miles per hour for 4 hours.

10 a snail moves 2 feet per hour for 3 hours.

5·3 Exercises

Evaluate each expression for the given value.

1. $3x - 11$ for $x = 9$
2. $3(6 - a) + 7 - 2a$ for $a = 4$
3. $(n \div 2) + 3n$ for $n = 6$
4. $3(2y - 1) + 6$ for $y = 2$

Use the formula $P = 2w + 2\ell$ to answer Exercises 5–8.

5. Find the perimeter of a rectangle that is 20 feet long and 15 feet wide.
6. Find the perimeter of the rectangle.

 7 cm

18 cm

7. Nia had a 20-inch by 30-inch enlargement made of a photograph. She wanted to have it framed. How many inches of frame would it take to enclose the photo?
8. The length of a standard football field is 100 yards. The width is 160 feet. What is the perimeter of a standard football field in feet? Remember to change the length to feet before finding the perimeter.

Use the formula $d = rt$ to answer Exercises 9–12.

9. Find the distance traveled by a walker who walks at 3 miles per hour for 2 hours.
10. A race car driver averaged 120 miles per hour. If the driver completed the race in $2\frac{1}{2}$ hours, how many miles was the race?
11. The speed of light is approximately 186,000 miles per second. About how far does light travel in 3 seconds?
12. During an electrical storm, you can use the speed of sound (1,100 feet per second) to estimate the distance of a storm. If you count the 8 seconds between the flash of lightning and the sound of thunder, about how far away was the lightning strike?

5·4 Equations

Equations

An **equation** is a mathematical sentence containing equivalent expressions separated by an equal sign. The equal sign means "is" or "is the same as." Equations may or may not contain variables. A few examples are shown below.

The sentence	Translates to
The difference of 14 and 5 is 9.	$14 - 5 = 9$
3 times a number is 6.	$3x = 6$
2 less than the product of a number and 5 is the same as 8.	$5x - 2 = 8$

Check It Out

Write an equation for each sentence.

1 9 subtracted from 5 times a number is 6.

2 6 added to a number is 10.

3 1 less than 4 times a number is 7.

True or False Equations

The equation $6 + 4 = 10$ represents a true statement. The equation $6 + 4 = 14$ is a false statement. Is the equation $x + 8 = 12$ true or false? You cannot determine whether it is true or false without knowing the value for x.

EXAMPLE Determine Whether the Equation Is True or False

Determine whether the equation $2x + 8 = 12$ is true or false for $x = 2$, $x = 4$, and $x = 6$.

$x + 8 = 12$	$x + 8 = 12$	$x + 8 = 12$
$2 + 8 \overset{?}{=} 12$	$4 + 8 \overset{?}{=} 12$	$6 + 8 \overset{?}{=} 12$
$10 \overset{?}{=} 12$	$12 \overset{?}{=} 12$	$14 \overset{?}{=} 12$
False	True	False

Determine whether each equation is true or false for
$x = 5$, $x = 6$, and $x = 7$.

④ $x + 3 = 8$

⑤ $4 + n = 10$

⑥ $25 - y = 18$

⑦ $p - 6 = 1$

Solve Equations Mentally

When you find the value for a variable that makes an equation a
true sentence, you *solve* the equation. The value for the variable
is the **solution** of the equation.

$$x + 3 = 7$$
$$4 + 3 = 7$$

This sentence is true. $\qquad 7 = 7$

The value for the variable x that makes the sentence true is 4,
so 4 is the solution.

EXAMPLE   Solve an Equation Mentally

Is 8, 9, or 10 the solution of the equation $b + 4 = 13$?

Value of b	$b + 4 \overset{?}{=} 13$	Are both sides equal?
8	$8 + 4 = 13$ $12 \neq 13$	no
9	$9 + 4 = 13$ $13 = 13$	yes
10	$10 + 4 = 13$ $14 \neq 13$	no

Because replacing b with 9 results in a true sentence, the
solution is 9.

Check It Out

Solve equations mentally.

8 Is 2, 3, or 5 the solution of the equation $6n = 18$?

9 Solve $36 \div t = 12$ mentally.

10 Evaluate $3a + b^2$ if $a = 2$ and $b = 3$.

Equivalent Equations

An *equivalent equation* can be obtained from an existing equation in one of four ways.

• Add the same term to both sides of the equation.
• Subtract the same term from both sides.
• Multiply by the same term on both sides.
• Divide by the same term on both sides.

Four equations equivalent to $x = 6$ are shown.

Operation	Equivalent Equation	Simplified
Add 3 to both sides.	$x + 3 = 6 + 3$	$x + 3 = 9$
Subtract 3 from both sides.	$x - 3 = 6 - 3$	$x - 3 = 3$
Multiply by 3 on both sides.	$x \times 3 = 6 \times 3$	$3x = 18$
Divide by 3 on both sides.	$\dfrac{x}{3} = \dfrac{6}{3}$	$\dfrac{x}{3} = 2$

Check It Out

Write equations equivalent to $x = 12$.

11 Add 3 to both sides.

12 Subtract 3 from both sides.

13 Multiply by 3 on both sides.

14 Divide by 3 on both sides.

Solving Equations by Addition or Subtraction

You can use equivalent equations to *solve* an equation.

Look at the equation $n + 6 = 9$. For the equation to be solved, n must be on one side of the equal sign by itself. How can you get rid of the $+ 6$ that is also on that side? You subtract 6. But remember that for the new equation to remain equivalent to $n + 6 = 9$, you must also subtract 6 from the other side.

$$n + 6 = 9$$
$$n + 6 - \mathbf{6} = 9 - \mathbf{6}$$
$$n = 3$$

You check the solution to be sure it is correct by substituting the solution for n and simplifying to see whether the sentence is true.

$$n + 6 = 9$$
$$3 + 6 \stackrel{?}{=} 9$$
$$9 = 9$$

Since the statement is true, $n = 3$ is the correct solution.

Equations involving subtraction, such as $b - 6 = 12$, can be solved similarly, except that you perform the operation of addition to both sides to get the variable by itself.

EXAMPLE Solving Equations by Addition

Solve the equation $b - 6 = 12$, and check your solution.

- Notice that for b to be on a side by itself, you need to remove the $- 6$.
- Add 6 to both sides.

$$b - 6 + \mathbf{6} = 12 + \mathbf{6}$$

- Simplify.

$$b = 18$$

- Check the solution. Substitute 18 for b in the equation.

$$18 - 6 \stackrel{?}{=} 12$$
$$12 = 12$$

The statement is true. $b = 18$ is the solution.

Solve each equation. Check your solution.

15 $y + 5 = 11$ **16** $n - 3 = 7$

17 $z + 9 = 44$ **18** $b - 6 = 2$

Solving Equations by Multiplication or Division

Look at the equation $3n = 6$. No number is being added to or subtracted from the variable. However, the n is still not by itself on one side of the equal sign. The n is being multiplied by 3.

To write an equivalent equation that isolates the n, you must perform the opposite operation. Division is the opposite of multiplication, so divide by 3 on both sides. Remember that a division expression can be written as a fraction.

$$3n = 6$$
$$3n \div 3 = 6 \div 3$$
$$\frac{3n}{3} = \frac{6}{3}$$
$$n = 2$$

Check the solution in the same way that you did with addition and subtraction. Substitute the possible solution for n. Simplify, and determine whether the sentence is true.

$$3n = 6$$
$$3(2) \overset{?}{=} 6$$
$$6 = 6$$

Because this is a true statement, $n = 2$ is the correct solution.

Equations involving division, such as $\frac{b}{3} = 4$, can be solved similarly, except that you perform the operation of multiplication to both sides to get the variable by itself.

EXAMPLE Solving Equations by Multiplication

Solve the equation $\frac{b}{3} = 4$, and check your solution.

- Notice that for b to be on a side by itself, you need to remove the denominator, 3.

$\frac{b}{3} \times 3 = 4 \times 3$ • Multiply both sides by 3.

$b = 12$ • Simplify.

$\frac{12}{3} \stackrel{?}{=} 4$ • Check the solution. Substitute 12 for b in the equation.

$4 = 4$

The statement is true. $b = 12$ is the solution.

➡️ **Check It Out**

Solve each equation. Check your solution.

19 $2x = 10$

20 $\frac{y}{3} = 8$

21 $4z = 16$

22 $\frac{a}{3} = 100$

Function Tables

A **function** is a relationship that describes how one number changes in relation to another number. It assigns exactly one output value to one input value. The input value is a variable—it can change. The number of the output value depends on the number of the input value.

A function rule describes the relationship between input and output values. The rule tells how the output will always change based on the input. The function rule can be shown in an equation. For example, in a function for which the rule says the output is always 2 more than the input, the equation is $f = x + 2$.

You can organize input and output values in a *function table*.

Substitute each input value for the variable x and solve to find the output value. Note that the input values do not need to be sequential. Any input value you choose to substitute in a function will result in one output value.

Input	Function Rule	Output
Values for the variable x	$x + 2$	Depends on the input value
10	$10 + 2$	12
11	$11 + 2$	13
12	$12 + 2$	14

EXAMPLE   Complete a Function Table

Complete a function table for this relationship: the output is 3 less than the input. Use 7, 8, and 9 as input values for the variable x.

- Write the function rule as an equation.

  $x - 3$

- Solve the equation using each value for x.

Input (x)	Function Rule $x - 3$	Output $(x - 3)$
10	$10 - 3$	7
11	$11 - 3$	8
12	$12 - 3$	9

Complete the function table.

Input (x)	Output $\left(\frac{x}{2}\right)$
10	5
12	6
20	10

Find the Rule for a Function Table

You can study the relationship between the inputs and outputs in a function table to find the function rule. Each output value in the table changes from its input value in the same way.

EXAMPLE Find the Rule for a Function Table

Find the rule for the function table.

Input (x)	Output (?)
3	6
4	8
5	10

$3 \times 2 = 6$ • Notice that each input is 2 times the output.

$4 \times 2 = 8$

$5 \times 2 = 10$

$2x$ • Write the function rule.

So, the rule for the function table is $2x$.

Check It Out

Complete each function table.

23

Input (x)	Output ($12x$)
2	
3	
4	

24

Input (x)	Output (?)
18	9
16	8
14	7

5·4 Exercises

Write an equation for each of the following:

1. 8 times a number is 36.

2. 4 less than 3 times a number is 20.

Determine whether each equation is true or false for $x = 7$, $x = 8$, and $x = 9$.

3. $x + 4 = 11$

4. $18 - n = 10$

5. Is 3, 5, or 7 the solution of the equation $3n = 21$?

6. Solve $63 \div t = 9$ mentally.

Solve each equation. Check your solution.

7. $y + 11 = 14$

8. $n - 3 = 77$

9. $z + 50 = 55$

10. $b - 7 = 3$

11. $4x = 12$

12. $\dfrac{y}{5} = 4$

13. $8z = 16$

14. $\dfrac{a}{11} = 10$

15. Complete the function table.

Input (x)	Output (x – 11)
12	
24	
36	

16. Find the rule for the function table.

Input (x)	Output (?)
3	34
8	39
13	44

5·5 Ratio and Proportion

Ratio

A **ratio** is a comparison of two quantities. If there are 10 boys and 15 girls in a class, the ratio of the number of boys to the number of girls is 10 to 15.

A ratio can be expressed as a fraction. The ratio of boys to girls is $\frac{10}{15}$, which reduces to $\frac{2}{3}$.

Ratios can also be written using a colon (:). The ratio of boys to girls is 2:3. You can make different comparisons and write other ratios.

Comparison	Ratio	As a Fraction
Number of girls to number of boys	15 to 10	$\frac{15}{10} = \frac{3}{2}$
Number of boys to number of students	10 to 25	$\frac{10}{25} = \frac{2}{5}$
Number of students to number of girls	25 to 15	$\frac{25}{15} = \frac{5}{3}$

Check It Out

A coin bank contains 8 dimes and 4 quarters. Write each ratio, and reduce to simplest form.

1 number of quarters to number of dimes
2 number of dimes to number of coins
3 number of coins to number of quarters

Rate

A **rate** is a ratio that compares two quantities that have different kinds of units. One example is a rate of travel, such as 90 miles in 3 hours. Miles and hours are different units.

The rate for one unit of a given quantity is called a **unit rate**. Some examples of unit rates are listed below.

$$\frac{55 \text{ mi}}{1 \text{ h}} \qquad \frac{5 \text{ apples}}{\$1} \qquad \frac{18 \text{ mi}}{1 \text{ gal}} \qquad \frac{\$400}{1 \text{ wk}} \qquad \frac{60 \text{ s}}{1 \text{ min}}$$

When written as a fraction, a unit rate always has a denominator of 1. So, to write a rate as a unit rate, divide the numerator and denominator of the rate by the denominator.

$$\frac{90 \text{ miles} \div 3}{3 \text{ hours} \div 3} = \frac{30 \text{ miles}}{1 \text{ hour}}$$

The unit rate is read as "30 miles per hour."

EXAMPLE Determining a Unit Rate

A booklet containing 10 movie admission passes costs $55.00. What is the unit rate?

$$\frac{\$55.00}{10 \text{ passes}}$$

- Write the comparison as a ratio.

$$\frac{\$55.00 \div 10}{10 \text{ passes} \div 10}$$

- Divide the numerator and denominator of the rate by the denominator.

$$\frac{\$5.50}{1 \text{ pass}}$$

The unit rate is $5.50 per pass.

➡️ **Check It Out**

Write each ratio as a unit rate.

④ $3.50 for 35 minutes

⑤ 108 points in six games

⑥ A rainforest can receive up to 90 centimeters of rain in 30 days.

⑦ An airplane climbs to an altitude of 1,000 feet in 10 minutes.

Proportions

When two ratios are equal, they form a **proportion**. If a car gets $\frac{27 \text{ mi}}{1 \text{ gal}}$, then the car can get $\frac{54 \text{ mi}}{2 \text{ gal}}$, $\frac{81 \text{ mi}}{3 \text{ gal}}$, and so on.

The ratios are all equal—they can be reduced to $\frac{27}{1}$.

$$\frac{54 \text{ mi} \div 2}{2 \text{ gal} \div 2} = \frac{27 \text{ mi}}{1 \text{ gal}} \qquad\qquad \frac{81 \text{ mi} \div 3}{3 \text{ gal} \div 3} = \frac{27 \text{ mi}}{1 \text{ gal}}$$

One way to determine whether two ratios form a proportion is to check their **cross products**. Every proportion has two cross products: the numerator of one ratio multiplied by the denominator of the other ratio. If the cross products are equal, the two ratios form a proportion.

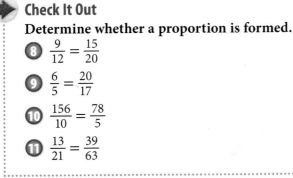

EXAMPLE   Determining a Proportion

Determine whether a proportion is formed by each pair of ratios.

$$\frac{6}{9} \overset{?}{=} \frac{45}{60} \qquad\qquad \frac{15}{9} \overset{?}{=} \frac{70}{42}$$

$$\frac{6}{9} \overset{?}{\diagdown} \frac{45}{60} \qquad\qquad \frac{15}{9} \overset{?}{\diagdown} \frac{70}{42}$$

$$6 \times 60 \overset{?}{=} 45 \times 9 \qquad 15 \times 42 \overset{?}{=} 70 \times 9$$

$$360 \neq 405 \qquad\qquad 630 = 630$$

This pair is not a proportion.   This pair is a proportion.

- Find the cross products.
- If the sides are equal, the ratios are proportional.

Check It Out

Determine whether a proportion is formed.

8 $\frac{9}{12} = \frac{15}{20}$

9 $\frac{6}{5} = \frac{20}{17}$

10 $\frac{156}{10} = \frac{78}{5}$

11 $\frac{13}{21} = \frac{39}{63}$

Using Proportions to Solve Problems

You can find an unknown value in a proportion by using the equivalent fraction method. Be sure that each ratio is set up in the same order.

Suppose that you can buy 3 pairs of socks for $4. This ratio is $\frac{3}{4}$. How much would it cost to buy 12 pairs of socks? Let c represent the unknown number, or variable—the cost of 12 pairs of socks. This ratio is $\frac{12}{c}$.

The two ratios are equal.

$$\overset{\times\ 4}{\underset{\times\ 4}{\underset{\text{dollars}}{\underset{\text{pairs of socks}}{\frac{3}{4}}} = \frac{12}{c}}}\ \begin{array}{l}\text{pairs of socks}\\\text{dollars}\end{array}$$

$$\frac{3}{4} = \frac{12}{16}$$

So, $c = 16$. 12 pairs of socks cost $16.

You can also use the cross products to solve for c. Because you have written a proportion, the cross products are equal.

$\frac{3}{4} = \frac{12}{c}$ $\qquad 3 \times c = 4 \times 12 \qquad 3c = 48$

To isolate c, divide both sides by 3.

$\frac{3c}{3} = \frac{48}{3}$

$c = 16$

Twelve pairs of socks cost $16.

Check It Out

Use proportions to solve.

12 A car gets 30 miles per gallon. How many gallons would the car need to travel 70 miles?

13 A worker earns $30 every 4 hours. How much would the worker earn in 14 hours?

14 The movie theater sells 5 tickets for $32.50. How much would 2 tickets cost?

5·5 Exercises

A basketball team has 10 wins and 5 losses. Write each ratio and reduce to lowest terms if possible.

1. number of wins to number of losses

2. number of wins to number of games

3. number of losses to number of games

Find the unit rate.

4. A swimming pool holds 21,000 gallons of water. It takes 28 hours to fill the pool with a garden hose. What is the unit rate at which the hose fills the pool?

5. It takes Casey from 3:00 to 5:00 P.M. to deliver 156 newspapers each afternoon. What is Casey's delivery rate?

6. Mariah downloaded 5 games to her cell phone for $16.25. What is the unit rate for the games?

Determine whether a proportion is formed.

7. $\dfrac{3}{5} = \dfrac{7}{11}$

8. $\dfrac{9}{6} = \dfrac{15}{10}$

9. $\dfrac{3}{4} = \dfrac{9}{16}$

Use a proportion to solve each problem.

10. In a class the ratio of boys to girls is $\dfrac{3}{2}$. If there are 12 boys in the class, how many girls are there?

11. A cell phone call costs $0.08 per minute. How much would a 6-minute call cost?

12. A map is drawn using a scale of 40 kilometers to 1 centimeter. The distance between two cities is 300 kilometers. How far apart are the two cities on the map?

13. A blueprint of a house is drawn using a scale of 5 meters to 2 centimeters. On the blueprint, a room is 6 centimeters long. How long will the actual room be?

14. The typical household in America uses an average of 5,328 kilowatt-hours (kWh) of electricity in 6 months. What is the average number of kilowatt-hours used in 4 months?

5·6 Inequalities

Showing Inequalities

When comparing the numbers 7 and 4, you might say that "7 is greater than 4" or that "4 is less than 7." When two expressions are not equal, or could be equal, you can write an **inequality**. The symbols are shown in the chart.

Symbol	Meaning	Example
>	is greater than	$7 > 4$
<	is less than	$4 < 7$
≥	is greater than or equal to	$x \geq 3$
≤	is less than or equal to	$-2 \leq x$

The equation $x = 5$ has one solution, 5. The inequality $x > 5$ has an infinite number of solutions: 5.001, 5.2, 6, 15, 197, and 955 are just some of the solutions. Note that 5 is not a solution because 5 is not greater than 5. You cannot list all of the solutions, but you can show them on a number line.

To show all the values that are greater than 5, but not including 5, use an open circle on 5 and shade the number line to the right.

Be sure to include an arrow at the end of your line to show that the solution set continues.

$$x > 5$$

The inequality $y \leq -1$ also has an infinite number of solutions: $-1.01, -1.5, -2, -8,$ and -54 are just some of the solutions. Note that -1 is also a solution because -1 is less than or equal to -1. On a number line, you want to show all the values that are less than or equal to -1. Because the -1 is to be included, use a closed (filled-in) circle on -1, and shade the number line to the left.

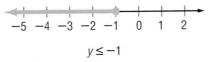

$y \leq -1$

Check It Out

Draw the number line showing the solutions to each inequality.

1 $x \geq 3$ **2** $y < -1$

3 $n > -2$ **4** $x \leq 0$

Solving Inequalities

Remember that addition and subtraction are opposite operations, as are multiplication and division. You can use opposite operations to solve inequalities. To solve the inequality $x + 4 > 7$, the x must be by itself on one side of the inequality symbol. Subtract 4 from both sides of the inequality.

$$x + 4 > 7$$
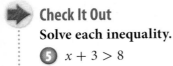
$$x + 4 - 4 > 7 - 4$$

$$x > 3$$

Inequality	Opposite Operation	Applied to Both Sides	Result
$n - 6 \leq 4$	add 6	$n - 6 + \leq 4 + 6$	$n \leq 10$
$n + 8 > 2$	subtract 8	$n + 8 - 8 > 2 - 8$	$n > -6$

Check It Out

Solve each inequality.

5 $x + 3 > 8$ **6** $y - 7 > 2$

5•6 Exercises

Choose a symbol from $<$, $>$, $\leq$, and $\geq$ for each blank.

1. 3 ___ 7

2. -8 ___ 4

3. 6 ___ 6

4. -2 ___ -7

Draw the number line showing the solutions to each inequality.

5. $x < 3$

6. $y \geq -1$

7. $n > 2$

8. $x \leq -4$

Solve each inequality.

9. $x - 4 < 5$

10. $2 + y \geq 8$

11. $n + 7 > 3$

12. $a - 3 \leq 6$

13. $3 + x \geq 15$

14. $x - 7 < 10$

15. $n - 8 \leq 1$

16. $5 + y > 6$

Choose the correct answer.

17. Which inequality has its solutions represented by this number line?

    A. $x < 3$ **B.** $x \leq 3$

    C. $x > 3$ **D.** $x \geq 3$

18. Which inequality would you solve by adding 5 to both sides?

    A. $x + 5 > 9$ **B.** $5x < 10$

    C. $\frac{x}{5} > 2$ **D.** $x - 5 < 4$

19. Which of the following statements is false?

    A. $-7 \leq 2$ **B.** $0 \leq -4$

    C. $8 \geq -8$ **D.** $5 \geq 5$

20. Which inequality would you graph by using an open circle and shading the number line to the right?

    A. $x < 6$ **B.** $x \leq 2$

    C. $x > -1$ **D.** $x \geq 2$

5·7 Graphing on the Coordinate Plane

Axes and Quadrants

When you cross a **horizontal** (left to right) number line with a **vertical** (up and down) number line, the result is a two-dimensional coordinate plane.

The number lines are called **axes**. The horizontal number line is the *x*-**axis**, and the vertical number line is the *y*-**axis**. The plane is divided into four regions, called **quadrants**. Each quadrant is named by a Roman numeral, as shown in the diagram.

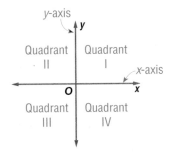

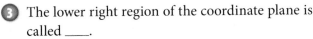

Check It Out

Fill in the blank.

1. The horizontal number line is called the ____.

2. The upper left region of the coordinate plane is called ____.

3. The lower right region of the coordinate plane is called ____.

4. the vertical number line is called the ____.

Writing an Ordered Pair

Any location on the coordinate plane can be represented by a
point. The location of any point is given in relation to where the
two axes intersect, called the **origin**.

Two numbers are required to
identify the location of a point.
The *x*-coordinate tells how far
to the left or right of the origin
the point lies. The *x*-coordinate
is always listed first.

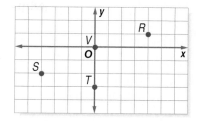

The *y*-coordinate tells how far up or down from the origin the
point lies. The *y*-coordinate is always listed second. Together,
the *x*-coordinate and *y*-coordinate form an **ordered pair**, (*x*, *y*).

Since point *R* is 4 units to the right of the origin and 1 unit up,
its ordered pair is (4, 1). Point *R* lies in Quadrant I. Point *S* is
4 units to the left of the origin and 2 units down, so its ordered
pair is (−4, −2). Point *S* lies in Quadrant III. Point *T* is 0 units
from the origin and 3 units down, so its ordered pair is (0, −3).
Point *T* lies on the *y*-axis. Point *V* is the origin, and its ordered
pair is (0, 0).

Check It Out

Give the ordered pair for each point.

5. *M*
6. *N*
7. *P*
8. *Q*

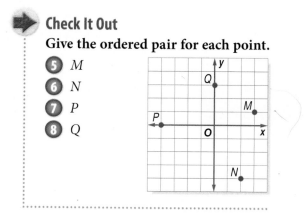

Locating Points on the Coordinate Plane

To locate point $A(3, -4)$, begin at
the origin and then move 3 units to the
right and 4 units down. Point A lies in
Quadrant IV. To locate point $B(-1, 4)$,
begin at the origin and then move 1 unit
to the left and 4 units up. Point B lies in
Quadrant II. Point $C(5, 0)$ is, from the
origin, 5 units to the right and 0 units up

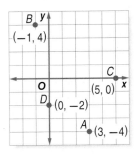

or down. Point C lies on the x-axis. Point $D(0, -2)$ is, from the
origin, 0 units to the left or right and 2 units down. Point D lies
on the y-axis.

Check It Out

**Locate each point on the coordinate plane and tell in which
quadrant or on which axis it lies.**

9 $H(4, -1)$ **10** $J(-1, 4)$ **11** $K(-2, -1)$ **12** $L(0, 2)$

The Graph of a Function Table

The input (x) and output (y) values in a function table are ordered
pairs and can be plotted on a graph. The function table below
shows four input and output values for the function rule $2x$.

The ordered pairs for this function table are $(1, 2)$, $(2, 4)$, $(3, 6)$,
and $(4, 8)$. The points are graphed on a coordinate plane.

Input (x)	Rule 2x	Output (2x)
1	2 · 1	2
2	2 · 2	4
3	2 · 3	6
4	2 · 4	8

Notice that the points lie along a straight line. The x- and
y-coordinates of any point on the line will result in a true
statement if they are substituted into the function equation $y = 2x$.

Complete a function table for this relationship: the output is 3 more than the input. Use 2, 3, 4, and 5 as input values for the variable x.

$y = x + 3$

- Write the function rule as an equation.

Input (x)	Rule $y = x + 3$	Output (y)
2	$y = 2 + 3$	5
3	$y = 3 + 3$	6
4	$y = 4 + 3$	7
5	$y = 5 + 3$	8

- Solve the equation using each value for x.

$(2, 5), (3, 6), (4, 7)$, and $(5, 8)$

- Write the input and output values as ordered pairs.

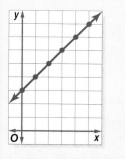

- Locate the points on a coordinate plane. Notice that the points lie along a straight line.

$(0, 3)$ or $(1, 4)$, for example

$4 = 1 + 3$ $3 = 0 + 3$

- Choose other points that lie on the line.
- Substitute the ordered pairs for the input and output values in the function equation $y = x + 3$.

$4 = 4$ $3 = 3$ These statements are true.

Check It Out

Create a function table and plot the points.

⓭ $y = x - 1$.
Use 5, 6, and 7 as input values.

Where's Your Antipode?

Imaginary lines of latitude and longitude cover the globe with a coordinate grid that is used to locate any place on Earth's surface. Latitude lines circle the globe in an east-west direction. They are drawn parallel to the equator starting at 0° and running to 90°N or 90°S at the poles.

Longitude lines circle the globe in a north-south direction, meeting at both poles. These lines run from 0° at the prime meridian to 180°E or 180°W, halfway around Earth. By using latitude and longitude as coordinates, you can pinpoint any location.

Antipodes (an-ti-pǝ-dēz) are places opposite each other on the globe. To find your antipode (an-tǝ-pōd), first find the latitude of your location and change the direction. For example, if your latitude is 56°N, the latitude of your antipode would be 56°S. Next find your longitude, subtract it from 180, and change the direction. For example, if your longitude is 120°E, the longitude of your antipode would be $180 - 120 = 60°$W.

Use a map to find the coordinates of your city. Find the coordinates of its antipode. Then locate your city and its antipode on a globe or map of the world.

5·7 Exercises

Fill in the blank.

1. The vertical number line is called the ____.

2. The lower left region of the coordinate plane is called ____.

3. The upper right region of the coordinate plane is called ____.

Give the ordered pair for each point.

4. *A*

5. *B*

6. *C*

7. *D*

8. *E*

9. *F*

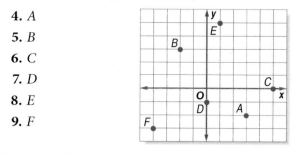

Locate each point on the coordinate plane and tell in which quadrant or on which axis it lies.

10. *H*(−3, −4)

11. *J*(0, 2)

12. *K*(4, −1)

13. *L*(−1, 0)

14. *M*(−3, 5)

15. *N*(3, 2)

16. Create a function table for $y = 6 - x$ with input values of 1, 2, 3, 4, and 5.

17. List the ordered pairs for Exercise 16. Plot the points on a coordinate plane.

Algebra

You can use the problems and the list of words that follow to see what you learned in this chapter. You can find out more about a particular problem or word by referring to the topic number (*for example,* Lesson 5·2).

Problem Set

Write an expression for each phrase. (Lesson 5·1)

1. a number decreased by 5

2. twice a number

3. 4 less than the sum of a number and 7

Simplify each expression. (Lesson 5·2)

4. $10x + 3 - 6x$

5. $8n + 3b - 5n - b$

6. Evaluate the expression $2x + 34$ for $x = 3$. (Lesson 5·3)

7. Write an equation for the following: 4 times a number is the same as 12 added to the same number. (Lesson 5·4)

8. Determine whether $126 \div p = 9$ is true or false for $p = 13$, $p = 14$, and $p = 15$. (Lesson 5·4)

9. Solve the equation $34x = 136$. Check your solution. (Lesson 5·4)

10. Find the distance traveled by a bicyclist who rides at 18 miles per hour for $2\frac{1}{2}$ hours. (Lesson 5·5)

Use a proportion to solve. (Lesson 5·5)

11. In a class, the ratio of boys to girls is $\frac{3}{4}$. If there are 24 girls in the class, how many boys are there?

12. A map is drawn using a scale of 120 kilometers to 1 centimeter. The distance between two cities is 900 kilometers. How far apart are the two cities on the map?

Solve each inequality. (Lesson 5·6)

13. $x - 4 \leq 2$ **14.** $4x > 12$

15. $n + 8 \geq 6$ **16.** $\frac{n}{4} < -1$

Locate each point on the coordinate plane and tell in which quadrant or on which axis it lies. (Lesson 5·7)

17. $A(-3, 2)$ **18.** $B(4, 0)$ **19.** $C(-2, -4)$

20. $D(3, 4)$ **21.** $E(0, -2)$ **22.** $F(-4, -1)$

Make a function table for each rule, using input values 2, 3, and 4. Plot the points for each on a coordinate plane. (Lesson 5·7)

23. $y = x - 3$ **24.** $y = 3x$

 **Write definitions for the following words.**

algebra (Lesson 5·1)
Associative Property (Lesson 5·2)
axes (Lesson 5·7)
Commutative Property (Lesson 5·2)
cross product (Lesson 5·5)
difference (Lesson 5·1)
Distributive Property (Lesson 5·2)
equation (Lesson 5·4)
equivalent (Lesson 5·4)
equivalent expression (Lesson 5·2)
expression (Lesson 5·1)
formula (Lesson 5·3)
function (Lesson 5·4)
horizontal (Lesson 5·7)
inequality (Lesson 5·6)
like terms (Lesson 5·2)

order of operations (Lesson 5·3)
ordered pair (Lesson 5·7)
origin (Lesson 5·7)
perimeter (Lesson 5·3)
point (Lesson 5·7)
product (Lesson 5·1)
proportion (Lesson 5·5)
quadrant (Lesson 5·7)
quotient (Lesson 5·1)
rate (Lesson 5·5)
ratio (Lesson 5·5)
solution (Lesson 5·4)
sum (Lesson 5·1)
term (Lesson 5·1)
unit rate (Lesson 5·5)
variable (Lesson 5·1)
vertical (Lesson 5·7)
x-axis (Lesson 5·7)
y-axis (Lesson 5·7)

HotTopic 6

Geometry

What do you know?

You can use the problems and the list of words that follow to see what you already know about this chapter. The answers to the problems are in **HotSolutions** at the back of the book, and the definitions of the words are in **HotWords** at the front of the book. You can find out more about a particular problem or word by referring to the topic number (*for example*, Lesson 6·2).

Problem Set

Use this figure for Exercises 1–3. (Lesson 6·1)

1. Name an angle.

2. Name a ray.

3. What is the measure of ∠*DBC*?

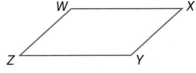

Use this figure for Exercises 4 and 5. (Lesson 6·2)

4. What kind of figure is quadrilateral *WXYZ*?

5. What is the sum of the measures of the angles of quadrilateral *WXYZ*?

6. What is the perimeter of a square that measures 4 feet on a side? (Lesson 6·4)

7. Find the area of a rectangle with length of 12 centimeters and width of 7 centimeters. (Lesson 6·5)

8. Each face of a pyramid is a triangle with a base of 8 inches and a height of 12 inches. If the area of its square base is 64 in^2, what is the surface area of the pyramid? (Lesson 6·6)

For Exercises 9–11, write the letter of the polyhedron, *A, B,* or *C,* to match it to its name. (Lesson 6·2)

9. triangular pyramid
10. cube
11. triangular prism

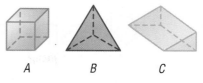

A   B   C

12. Find the volume of a rectangular prism whose base measures 6 cm by 2 cm and whose height is 5 cm. (Lesson 6·7)

Use circle *T* to answer Exercises 13 and 14. (Lesson 6·8)

13. What is the diameter of circle *T*?
14. What is the area of circle *T* rounded to the nearest tenth of an inch?

T 3 in.

HotWords

adjacent angles (Lesson 6·1)
angle (Lesson 6·1)
circumference (Lesson 6·6)
complementary angles
    (Lesson 6·1)
congruent angles (Lesson 6·1)
cube (Lesson 6·2)
degree (Lesson 6·1)
diagonal (Lesson 6·2)
diameter (Lesson 6·8)
face (Lesson 6·2)
net (Lesson 6·6)
opposite angle (Lesson 6·2)
parallelogram (Lesson 6·2)
perimeter (Lesson 6·4)
pi (Lesson 6·8)
point (Lesson 6·2)
polygon (Lesson 6·1)
polyhedron (Lesson 6·1)
prism (Lesson 6·2)

pyramid (Lesson 6·2)
quadrilateral (Lesson 6·2)
radius (Lesson 6·8)
ray (Lesson 6·1)
rectangular prism (Lesson 6·2)
reflection (Lesson 6·3)
regular polygon (Lesson 6·2)
right angle (Lesson 6·1)
rotation (Lesson 6·3)
supplementary angles
    (Lesson 6·1)
surface area (Lesson 6·5)
symmetry (Lesson 6·3)
transformation (Lesson 6·3)
translation (Lesson 6·3)
triangular prism (Lesson 6·6)
vertex (Lesson 6·1)
vertical angles (Lesson 6·1)
volume (Lesson 6·7)

6·1 Naming and Classifying Angles and Triangles

Points, Lines, and Rays

In mathematics, it is sometimes necessary to refer to a specific **point** in space. A point has no size; its only function is to show position. You show a point as a small dot and name it with a single capital letter.

$$\bullet \\ A$$

Point *A*

If you draw two points on a sheet of paper, a *line* can be used to connect them. Imagine this line as being perfectly straight and continuing without end in opposite directions. It has no thickness. To name a line, choose any two points on the line.

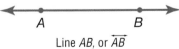

Line *AB*, or $\overleftrightarrow{AB}$

A **ray** is part of a line that extends without end in one direction. In $\overrightarrow{AB}$, which is read as "ray *AB*," *A* is the endpoint. The second point that is used to name the ray can be any point other than the endpoint. You could also name this ray *AC*.

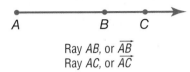

Ray *AB*, or $\overrightarrow{AB}$
Ray *AC*, or $\overrightarrow{AC}$

Use symbols to name each line or ray in two ways.

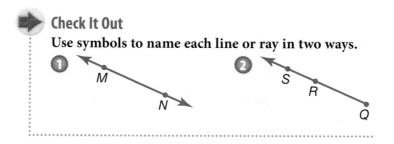

Naming Angles

Imagine two different rays with the same endpoint. Together they form an **angle**. The point they have in common is called the **vertex** of the angle. The rays form the *sides* of the angle.

The angle above is made up of $\overrightarrow{BA}$ and $\overrightarrow{BC}$. B is the common endpoint of the two rays. Point B is the vertex of the angle. Instead of writing the word *angle*, you can use the symbol for angle $\angle$.

You can name an angle using the three letters of the points that make up the two rays with the vertex as the middle letter ($\angle ABC$ or $\angle CBA$). You can also use just the letter of the vertex to name the angle ($\angle B$). Sometimes, you may want to name an angle with a number ($\angle 2$).

When more than one angle is formed at a vertex, you use three letters to name each of the angles. Because G is the vertex of three different angles, each angle needs three letters to name it: $\angle DGF$ or $\angle FGD$, $\angle DGE$ or $\angle EGD$, $\angle EGF$ or $\angle FGE$.

Check It Out
Find three angles in the figure below.

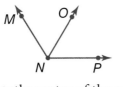

③ Name the vertex of the angles.

④ Use symbols to name each angle.

Measuring Angles

You measure an angle in **degrees** by using a protractor. The number of degrees in an angle will be greater than 0 and less than or equal to 360. The notation $m\angle$ is used to denote the measure of an angle.

EXAMPLE   **Measuring with a Protractor**

Find $m\angle ABC$ and $m\angle MNO$.

- Place the center point of the protractor over the vertex of the angle. Line up the 0° line on the protractor with one side of the angle.
- Find the point at which the other side of the angle crosses the protractor. Read the number of degrees on the scale at that point.
- Measure $\angle ABC$ and $\angle MNO$.

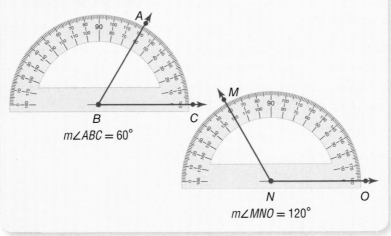

$m\angle ABC = 60°$

$m\angle MNO = 120°$

Check It Out

Use a protractor to measure each angle.

5 ∠*EFG*

6 ∠*PFR*

7 ∠*GFQ*

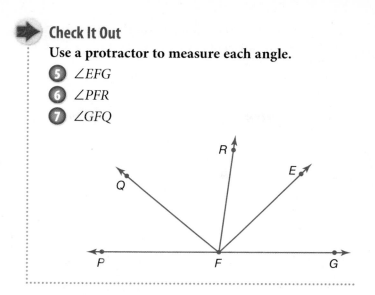

Classifying Angles

You can classify angles by their measures.

Acute angle
measures less than 90°

Right angle
measures 90°

Obtuse angle
measures greater than 90°
and less than 180°

Straight angle
measures 180°

Reflex angle
measures greater than 180°

Angles that share a side are called **adjacent angles**. You can add measures if the angles are adjacent.

$m\angle APB = 55°$

$m\angle BPC = 35°$

$m\angle APC = 55° + 35° = 90°$

Because the sum is 90°, you know that $\angle APC$ is a **right angle**.

Check It Out

Use a protractor to measure the angle, and then classify it.

8 $\angle KST$

9 $\angle RST$

10 $\angle FST$

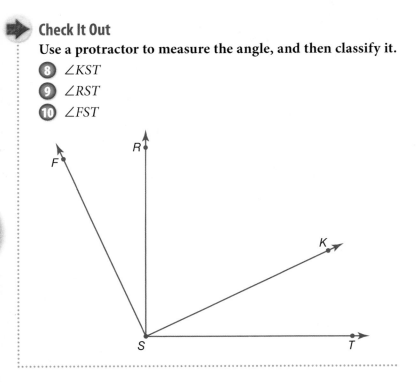

Ancient Egyptian obelisks were carved horizontally out of a rock quarry. Exactly how the Egyptians lifted the obelisks into vertical position is a mystery. But clues suggest that the Egyptians slid them down a dirt ramp, inched them higher with levers, and finally pulled them upright with ropes.

A crew from a television station attempted to move a 43-foot-long block of granite using this method. They tilted the 40-ton obelisk down a ramp at a 33° angle. With levers, they inched the obelisk up to about a 40° angle. Then 200 people tried to haul it with ropes to a standing position. They couldn't budge it. Finally, out of time and money, they abandoned the attempt.

How many additional degrees did the crew need to raise the obelisk before it would have stood upright? See **HotSolutions** for the answer.

Special Pairs of Angles

When two lines intersect, they form two pairs of opposite angles. The opposite angles are called **vertical angles**, and they have the same measure. Angles with the same measure are called **congruent angles**. The symbol ≅ is used to show congruence.

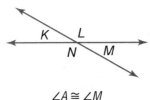

$$\angle A \cong \angle M$$
$$\angle L \cong \angle N$$

Two angles that have a sum of 180° are **supplementary angles**.

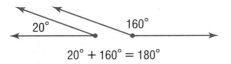

$$20° + 160° = 180°$$

Two angles that have a sum of 90° are **complementary angles**.

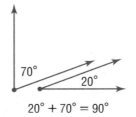

$$20° + 70° = 90°$$

Check It Out

Classify the following pairs of angles as *supplementary,* *complementary,* or *vertical,* and tell whether the pair is congruent.

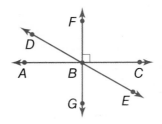

11 ∠CBE and ∠EBG
12 ∠DBF and ∠EBG
13 ∠ABD and ∠DBC

Triangles

Triangles are *polygons* (p. 267) that have three sides, three vertices, and three angles.

Triangles are named using their three vertices in any order. $\triangle ABC$ is read "triangle ABC."

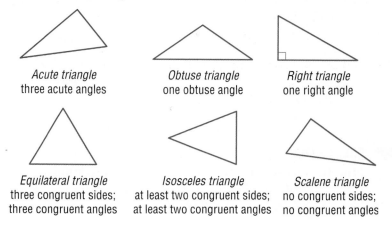

Acute triangle three acute angles	*Obtuse triangle* one obtuse angle	*Right triangle* one right angle
Equilateral triangle three congruent sides; three congruent angles	*Isosceles triangle* at least two congruent sides; at least two congruent angles	*Scalene triangle* no congruent sides; no congruent angles

Classifying Triangles

Like angles, triangles are classified by their angle measures. They are also classified by the number of *congruent* sides.

The sum of the measures of the three angles in a triangle is always 180°.

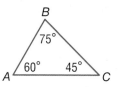

In $\triangle ABC$, $m\angle A = 60°$, $m\angle B = 75°$, and $m\angle C = 45°$.

$60° + 75° + 45° = 180°$

So, the sum of the angles of $\triangle ABC$ is 180°.

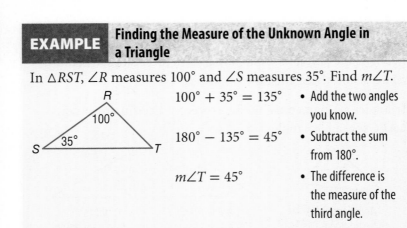

EXAMPLE Finding the Measure of the Unknown Angle in a Triangle

In △RST, ∠R measures 100° and ∠S measures 35°. Find $m\angle T$.

$100° + 35° = 135°$ • Add the two angles you know.

$180° - 135° = 45°$ • Subtract the sum from 180°.

$m\angle T = 45°$ • The difference is the measure of the third angle.

So, $m\angle T$ is 45°.

Check It Out

Find the measure of the third angle of each triangle.

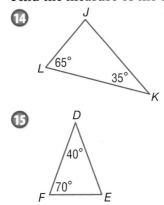

14

15

6•1 Exercises

Use symbols to name each line in two ways.

1.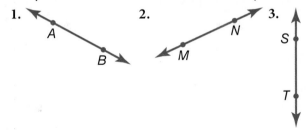

2.

3.

Name each ray in two ways.

4.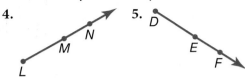

5.

6.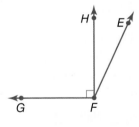

Find the three angles in this figure.

7. Name the vertex of each angle.

8. Name the acute angle.

9. Name the right angle.

10. Name the obtuse angle.

Use this figure for Exercises 11–13.

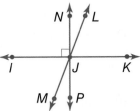

11. Name the pairs of complementary angles.

12. How many pairs of vertical angles are shown?

13. How many pairs of supplementary angles are shown? (Hint: Don't forget to include the pairs of right angles!)

Use this figure for Exercises 14–16.

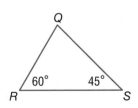

14. What is the measure of ∠Q?

15. What kind of triangle is △QRS: acute, right, or obtuse?

16. Classify ∠Q.

6·2 Polygons and Polyhedrons

Quadrilaterals

A **quadrilateral** is a figure with four sides and four angles. There are many different types of quadrilaterals, which are classified by their sides and angles.

To name a quadrilateral, list the four vertices, either clockwise or counterclockwise. The quadrilateral at the right can be named *FISH*.

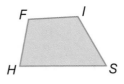

Angles of a Quadrilateral

The sum of the angles of a quadrilateral is 360°. If you know the measures of three angles of a quadrilateral, you can find the measure of the fourth angle.

EXAMPLE	Finding the Measure of the Unknown Angle in a Quadrilateral

Find the measure of ∠S in quadrilateral *STUV*.

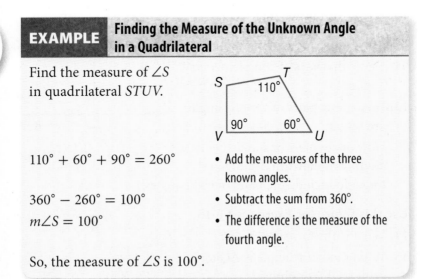

$110° + 60° + 90° = 260°$ • Add the measures of the three known angles.

$360° - 260° = 100°$ • Subtract the sum from 360°.

$m∠S = 100°$ • The difference is the measure of the fourth angle.

So, the measure of ∠S is 100°.

Check It Out

Use the quadrilateral to answer the questions.

1 Name the quadrilateral in two ways.

2 What is the sum of the measures of ∠M, ∠N, and ∠O?

3 Find m∠L.

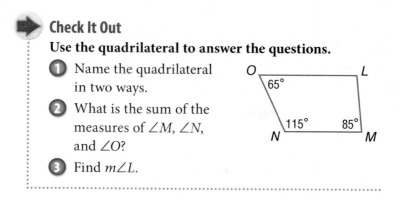

Types of Quadrilaterals

A *rectangle* is a quadrilateral with four right angles. *WXYZ* is a rectangle. Its length is 5 cm and its width is 3 cm.

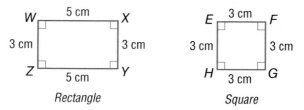

Rectangle   Square

Opposite sides of a rectangle are equal in length. If all four sides of the rectangle are equal, the rectangle is called a *square*. A square is a *regular polygon* because all of its sides are of equal length and all of the interior angles are *congruent*. Some rectangles may be squares, but *all* squares are rectangles. So, *EFGH* is both a square and a rectangle.

A **parallelogram** is a quadrilateral with opposite sides that are *parallel*. In a parallelogram, opposite sides are equal, and **opposite angles** are equal. *ABCD* is a parallelogram. *HIJK* is both a parallelogram and a rectangle.

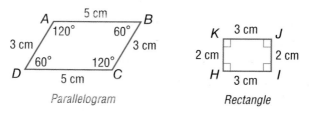

Parallelogram   Rectangle

Some parallelograms may be rectangles, but *all* rectangles are parallelograms. Therefore, squares are also parallelograms. If all four sides of a parallelogram are the same length, the parallelogram is called a *rhombus*. *HIJK* is a rhombus.

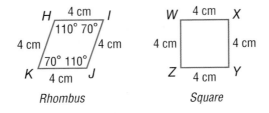

Rhombus   Square

Every square is a rhombus, although not every rhombus is a square, because a square also has equal angles.

In a *trapezoid,* two sides are parallel and two are not. A trapezoid is a quadrilateral, but it is not a parallelogram. *PARK* is a trapezoid.

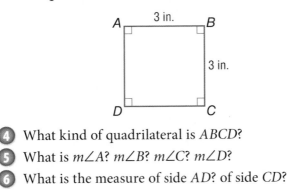

6•2

➡️ **Check It Out**

Use the quadrilateral to answer Exercises 4–6.

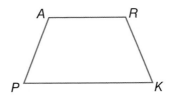

4 What kind of quadrilateral is *ABCD*?

5 What is $m\angle A$? $m\angle B$? $m\angle C$? $m\angle D$?

6 What is the measure of side *AD*? of side *CD*?

Polygons

A **polygon** is a closed figure that has three or more sides. Each side is a line segment, and the sides meet only at the endpoints, or vertices.

This figure is a polygon.   These figures are not polygons.

A rectangle, a square, a parallelogram, a rhombus, a trapezoid, and a triangle are all polygons.

Some aspects of polygons are always true. For example, a polygon of n sides has n angles and n vertices; a polygon with three sides has three angles and three vertices. A polygon with eight sides has eight angles and eight vertices, and so on.

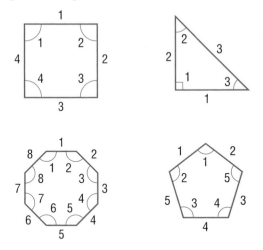

A **regular polygon** is a polygon in which all sides are of equal length and all interior angles are congruent. Three of the four shapes shown above are regular polygons. The triangle is not a regular polygon; its three sides are not of equal length.

A line segment connecting two vertices of a polygon is either a side or a **diagonal**. $\overline{AE}$ is a side of polygon $ABCDE$. $\overline{AD}$ is a diagonal.

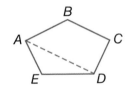

Types of Polygons

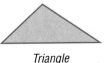

Triangle	Quadrilateral	Pentagon	Hexagon	Octagon
3 sides	4 sides	5 sides	6 sides	8 sides

A seven-sided polygon is called a *heptagon,* a nine-sided polygon is called a *nonagon,* and a ten-sided polygon is called a *decagon.*

Check It Out

Name each polygon.

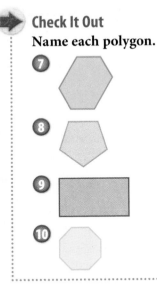

7

8

9

10

Polyhedrons

Solid shapes can be curved, like these.

| Sphere | Cylinder | Cone |

Some solid shapes have all flat surfaces. Each of the figures below is a *polyhedron*.

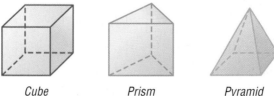

| Cube | Prism | Pyramid |

A **polyhedron** is any solid whose surface is made up of polygons. Triangles, quadrilaterals, and pentagons make up the **faces** of the common polyhedrons below.

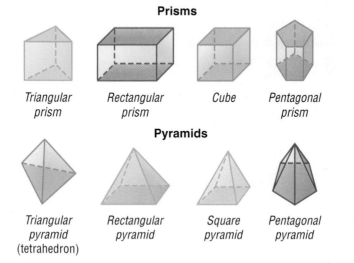

Prisms

| Triangular prism | Rectangular prism | Cube | Pentagonal prism |

Pyramids

| Triangular pyramid (tetrahedron) | Rectangular pyramid | Square pyramid | Pentagonal pyramid |

A **prism** has two bases, or "end" faces. The *bases* of a prism are the same size and shape and are parallel to each other. The prism's other faces are parallelograms. When all six faces of a **rectangular prism** are square, the figure is a **cube**.

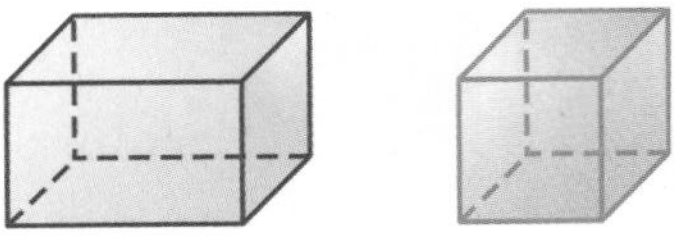

Rectangular prism *Cube*

A **pyramid** is a structure that has one base in the shape of a polygon. The pyramid's triangular faces meet at a point called the *apex*. The base of each pyramid shown below is shaded.

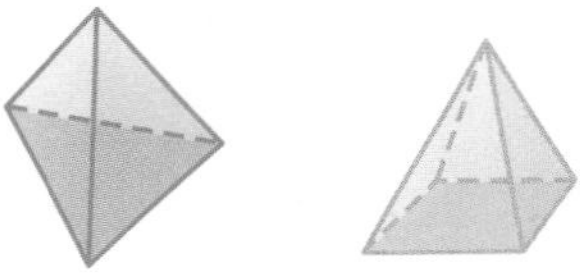

Triangular pyramid (tetrahedron) *Square pyramid*

Check It Out

Identify each polyhedron.

11 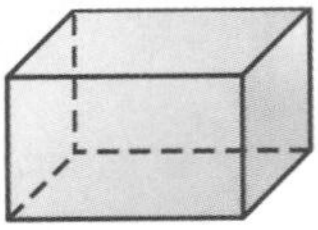

12 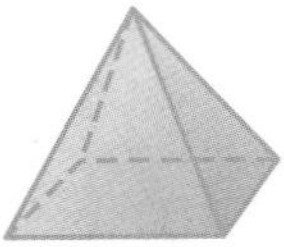

6·2 Exercises

Use the figures to the right for Exercises 1–4.

1. Find $m\angle J$.
2. Give two other names for quadrilateral *IJKL*.

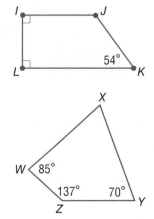

3. Find $m\angle X$.
4. Give two other names for quadrilateral *WXYZ*.

Write the letter of the word in the word box that matches each description.

5. It is a solid figure whose surface is made up of polygons.
6. Its two pairs of opposite sides and opposite angles are equal.
7. It is a quadrilateral with sides of equal length.
8. It is a quadrilateral with opposite sides parallel and opposite angles equal.
9. It is a closed figure that has three or more sides.
10. It is a rectangular prism with six square faces.
11. It is a quadrilateral with four right angles and sides of two different lengths.

Word Box
A. rhombus
B. polyhedron
C. rectangle
D. square
E. cube
F. polygon
G. parallelogram

Use quadrilateral *EFGH* for Exercises 12 and 13.

12. Give two other names for quadrilateral *EFGH*.
13. Find the measure of $\angle G$.

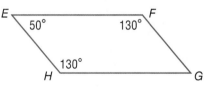

Identify each polygon.

14.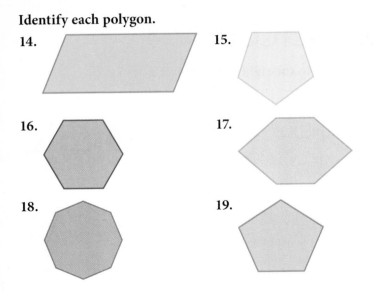

15.

16.

17.

18.

19.

Two angles of △XYZ measure 23° and 67°.

20. What is the measure of the third angle?

21. What kind of triangle is △XYZ?

22. Which of the following shapes are regular polygons?

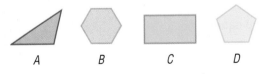

A   B   C   D

23. Name the polyhedron that has two identical bases that are parallel pentagons.

24. Name the polyhedron that has a square base and faces that are triangles.

Identify the shapes of these real-world polyhedrons.

25.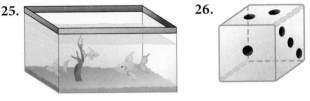

Fish tank

26.

6·3 Symmetry and Transformations

Whenever you move a shape that is in a plane, you are performing a **transformation**.

Reflections

A **reflection** (or flip) is one kind of transformation. When you hear the word "reflection," you may think of a mirror. The mirror image, or reverse image, of a point or shape is called a *reflection*.

The reflection of a point is another point on the opposite side of a line of **symmetry**. Both the point and its reflection are the same distance from the line.

P' reflects point P on the other side of line ℓ. P' is read "P-prime." P' is called the *image* of P.

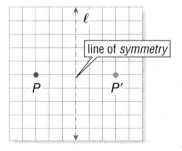

Any point, line, or polygon can be reflected. Quadrilateral *DEFG* is reflected on the other side of line *m*. The image of *DEFG* is *D'E'F'G'*.

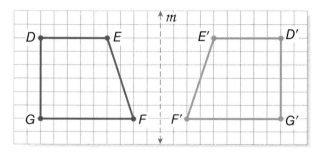

In the quadrilateral reflection on the opposite page, point *D* is 10 units from the line of symmetry, and point *D'* is also 10 units from the line on the opposite side. You can measure the distance from the line for each point, and the corresponding image point will be the same distance.

Check It Out

Draw the image.

1. Draw figure *ABCD* and line *e* on grid paper. Then draw and label the reflection of the figure on line *e*.

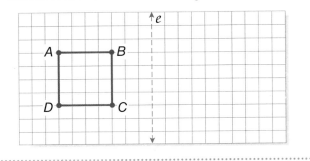

Reflection Symmetry

You have seen that a line of symmetry is used to show the reflection symmetry of a point, a line, or a shape. A line of symmetry can also *separate* a single shape into two parts, where one part is a reflection of the other. Each of these figures is symmetrical with respect to the line of symmetry.

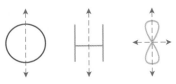

Sometimes a figure has more than one line of symmetry. Here are three shapes that have more than one line of symmetry.

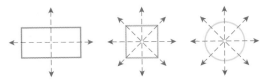

Some capital letters have reflection symmetry.

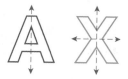

2 The letter A has one line of symmetry. Which other capital letters have exactly one line of symmetry?

3 The letter X has two lines of symmetry. Which other capital letters have two lines of symmetry?

Rotations

A **rotation** (or turn) is a transformation that turns a line or a shape around a fixed point. This point is called the *center of rotation*. Degrees of rotation are usually measured in the counterclockwise direction.

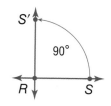

$\overrightarrow{RS}$ is rotated 90° around point *R*.

If you rotate a figure 360°, it returns to where it started. Despite the rotation, its position is unchanged. If you rotate $\overrightarrow{AB}$ 360° around point *P*, $\overrightarrow{AB}$ is still in the same place.

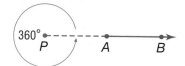

Check It Out

How many degrees has the flag been rotated around point H or J?

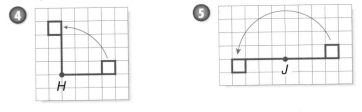

Translations

A **translation** (or slide) is another kind of transformation. When you slide a figure to a new position without turning it, you are performing a translation.

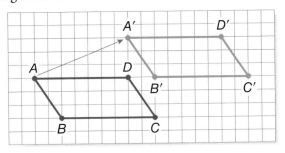

$A'B'C'D'$ is the image of $ABCD$ under a translation. A' is 9 units to the right and 4 units up from A. All other points on the rectangle translate in the exact same way.

Check It Out

Use the illustration to answer the question.

6 Which figures are translations of the shaded figure?

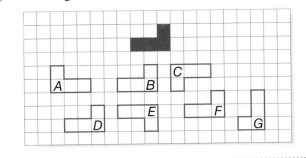

6·3 Exercises

What is the reflected image across line *s* of each of the following?

1. Point *D*   **2.** $\overline{DF}$ **3.** △*DEF*

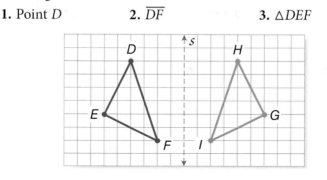

Trace each shape. Then draw all lines of symmetry.

4.   **5.**   **6.**

For each transformation, tell whether the image shows a reflection, a rotation, or a translation.

7.   **8.**

9.   **10.**

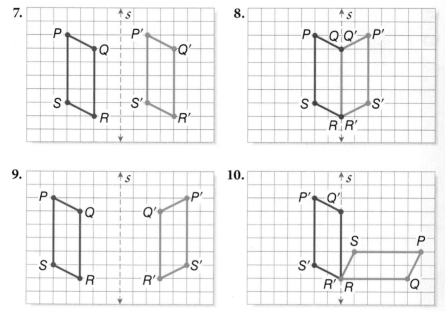

6·4 Perimeter

Perimeter of a Polygon

Ramon is planning to put a fence around his pasture. To determine how much fencing he needs, he must calculate the **perimeter** (P), or distance around, of his pasture.

Ramon's Pasture

180 yd

120 yd   120 yd

150 yd

The perimeter of any polygon is the sum of the lengths of the sides of the polygon. To find the perimeter of his pasture, Ramon needs to measure the length of each side and to find the sum. He finds that two of the sides are 120 yards, one side is 180 yards, and another is 150 yards. How much fencing will Ramon need to enclose his pasture?

$$P = 120 \text{ yd} + 150 \text{ yd} + 120 \text{ yd} + 180 \text{ yd} = 570 \text{ yd}$$

Ramon will need 570 yards of fencing to enclose the pasture.

EXAMPLE   **Finding the Perimeter of a Polygon**

Find the perimeter of the hexagon.

- Add the lengths of its sides.

$P = 5 + 10 + 8 + 10 + 5 + 18 = 56 \text{ ft}$
The perimeter is 56 feet.

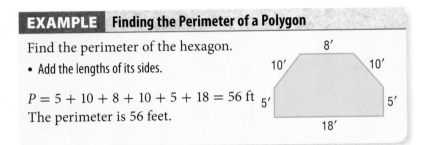

8'

10'   10'

5'   5'

18'

Regular Polygon Perimeters

The sides of a regular polygon are all the same length. If you know the perimeter of a regular polygon, you can find the length of each side.

To find the length of each side of a regular octagon with a perimeter of 36 cm, let $x =$ length of a side.

$$36 \text{ cm} = 8x$$
$$4.5 \text{ cm} = x$$

Each side is 4.5 centimeters long.

Perimeter of a Rectangle

Opposite sides of a rectangle are equal in length. So, to find the perimeter of a rectangle, you need to know only its length and width. The formula for the perimeter of a rectangle is $P = 2\ell + 2w$.

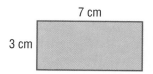

The perimeter of this rectangle is
$$(2 \times 7 \text{ cm}) + (2 \times 3 \text{ cm}) = 20 \text{ cm}.$$

Check It Out

Find the perimeter of each polygon.

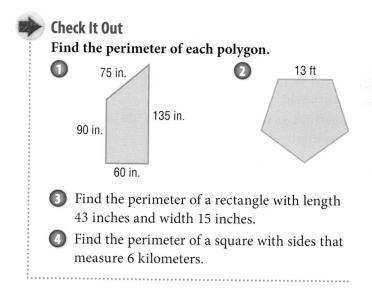

1 75 in. 135 in. 90 in. 60 in.

2 13 ft

3 Find the perimeter of a rectangle with length 43 inches and width 15 inches.

4 Find the perimeter of a square with sides that measure 6 kilometers.

6•4 Exercises

Find the perimeter of each polygon.

1.

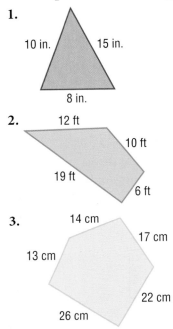

10 in. 15 in.

8 in.

2. 12 ft

10 ft

19 ft

6 ft

3. 14 cm

17 cm

13 cm

22 cm

26 cm

4. Find the perimeter of a square that measures 19 meters on a side.

5. Find the perimeter of a regular hexagon that measures 6 centimeters on a side.

6. The perimeter of a regular pentagon is 140 inches. Find the length of each side.

7. The perimeter of an equilateral triangle is 108 millimeters. What is the length of each side?

Find the perimeter of each rectangle.

8. $\ell = 28$ m, $w = 11$ m

9. $\ell = 25$ yd, $w = 16$ yd

10. $\ell = 43$ ft, $w = 7$ ft

6·5 Area

What Is Area?

Area measures the size of a surface. Your desktop is a surface with area, and so is the state of Montana. Instead of measuring with units of length, such as inches, centimeters, feet, and kilometers, you measure area in square units, such as *square inches* (in²) and *square centimeters* (cm²).

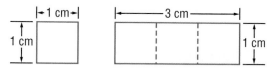

The square has an area of one square centimeter. It takes exactly three of these squares to cover the rectangle, which tells you that the area of the rectangle is three square centimeters, or 3 cm².

Estimating Area

When an exact answer is not needed or is difficult to find, you can *estimate* the area of a surface.

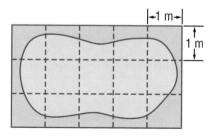

In the region shaded blue, three squares are completely shaded, so you know that the area is greater than 3 m². The area of the entire rectangle around the shape is 15 m², so the shaded area is less than 15 m². You can estimate the area of the shaded region by saying that it is greater than 3 m² but less than 15 m².

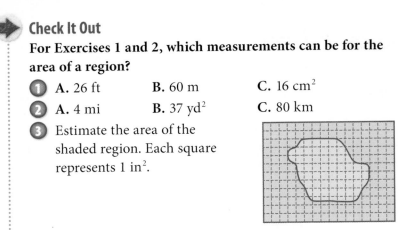

Check It Out

For Exercises 1 and 2, which measurements can be for the area of a region?

1 **A.** 26 ft   **B.** 60 m   **C.** 16 cm^2

2 **A.** 4 mi   **B.** 37 yd^2 **C.** 80 km

3 Estimate the area of the shaded region. Each square represents 1 in^2.

Area of a Rectangle

You can count squares to find the area of this rectangle.

Each of the 24 squares measures a square centimeter. So, the area of this rectangle is 24 cm^2.

You can also use the formula for finding the area of a rectangle:

$A = \ell \times w$.

The length of the rectangle above is 6 centimeters and the width is 4 centimeters. Using the formula, you find that

$A = 6 \text{ cm} \times 4 \text{ cm} = 24 \text{ cm}^2$.

AREA

6·5

If the rectangle is a square, the length and the width are the same. So for a square with side measure s units, you can use the formula $A = s \times s$, or $A = s^2$.

EXAMPLE   **Finding the Area of a Rectangle**

Find the area of this rectangle.

3 ft

16 in.

- The length and the width must be expressed in terms of the same units.
    3 ft = 36 in. So, ℓ = 36 in. and w = 16 in.
- Use the formula for the area of a rectangle: $A = \ell \times w$.
    $A = 36$ in. $\times$ 16 in.
- Multiply and solve.
    $A = 576$ in^2

The area of the rectangle is 576 in^2.

Check It Out

Find the area.

4 Find the area of a rectangle with a length of 24 yards and a width of 17 yards.

5 Find the area of a square that measures 9 inches on a side.

Area of a Parallelogram

To find the area of a parallelogram, multiply the base by the height.
Area = base × height
$A = b \times h$,
or $A = bh$

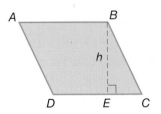

The height of a parallelogram is always perpendicular to the base. So, in parallelogram $ABCD$, the height, h, is equal to the length of $\overline{BE}$, not the length of $\overline{BC}$. The base, b, is equal to the length of $\overline{DC}$.

EXAMPLE **Finding the Area of a Parallelogram**

Find the area of a parallelogram with a base of 12 inches and a height of 7 inches.

$A = 12$ in. $\times$ 7 in.

- Use the formula: $A = b \times h$.

$A = 84$ in^2

- Solve.

The area of the parallelogram is 84 in^2.

Check It Out

Solve.

6 Find the area of a parallelogram with a base of 12 centimeters and a height of 15 centimeters.

7 Find the length of the base of a parallelogram whose area is 56 ft^2 and whose height is 7 feet.

APPLICATION **Fish Farming**

Fish farming, or *aquaculture*, is one of the fastest-growing food industries and now supplies about 20 percent of the world's fish and shellfish.

An oyster farmer builds large floating rafts in the ocean, and then hangs clean shells from them by ropes. Oyster larvae attach to the shells and grow in thick masses. The rafts are supported with barrels so that they don't sink to the bottom, where the oysters' natural predators can reach them.

An oyster farmer might have 100 rafts, each about 10 meters by 15 meters. What is the total area of these rafts? See **HotSolutions** for the answer.

Area of a Triangle

If you cut a parallelogram along a diagonal, you would have two triangles with equal bases, b, and the same height, h.

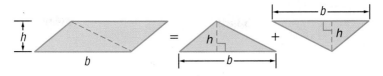

A triangle has half the area of a parallelogram with the same base and height. The area of a triangle equals $\frac{1}{2}$ the base times the height, so the formula is $A = \frac{1}{2} \times b \times h$, or $A = \frac{1}{2} bh$.

$A = \frac{1}{2} \times b \times h$

$A = \frac{1}{2} \times 13.5 \text{ cm} \times 8.4 \text{ cm}$

$\quad = 0.5 \times 13.5 \text{ cm} \times 8.4 \text{ cm}$

$\quad = 56.7 \text{ cm}^2$

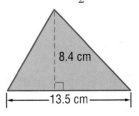

The area of the triangle is 56.7 cm².

EXAMPLE | **Finding the Area of a Triangle**

Find the area of $\triangle PQR$. Note that in a right triangle the two legs serve as a height and a base.

$A = \frac{1}{2} \times 5 \text{ m} \times 3 \text{ m}$ • Use the formula
$\qquad\qquad\qquad\qquad A = \frac{1}{2} bh.$

$\quad = 0.5 \times 5 \text{ m} \times 3 \text{ m}$ • Multiply.

$\quad = 7.5 \text{ m}^2$ • Solve.

The area of the triangle is 7.5 m².

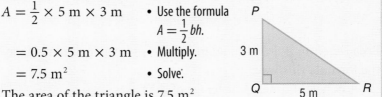

Check It Out

Solve.

8 Find the area of a triangle whose base is 16 feet and whose height is 8 feet.

9 Find the area of a right triangle whose sides measure 6 centimeters, 8 centimeters, and 10 centimeters.

Area of a Trapezoid

A trapezoid has two bases labeled b_1 and b_2. You read b_1 as "b sub-one." The area of a trapezoid is equal to the area of two triangles.

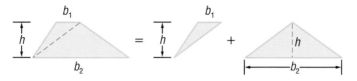

You know that the formula for the area of a triangle is $A = \frac{1}{2}bh$, so it makes sense that the formula for finding the area of a trapezoid would be $A = \frac{1}{2}b_1h + \frac{1}{2}b_2h$, or in simplified form, $A = \frac{1}{2}h(b_1 + b_2)$.

EXAMPLE   Finding the Area of a Trapezoid

Find the area of trapezoid $WXYZ$.

$A = \frac{1}{2} \times 4(5 + 11)$ • Use the formula
$\qquad\qquad\qquad\qquad A = \frac{1}{2}h(b_1 + b_2)$.

$\quad = 2 \times 16$ • Multiply.

$\quad = 32 \text{ cm}^2$ • Solve.

The area of the trapezoid is 32 cm^2.

$W \quad b_1 = 5 \text{ cm} \quad X$
$h = 4 \text{ cm}$
$Z \quad b_2 = 11 \text{ cm} \quad Y$

Because $\frac{1}{2}h(b_1 + b_2)$ is equal to $h \times \dfrac{b_1 + b_2}{2}$, you can remember the formula this way:

$\quad A = $ height times the average of the bases.

For a review of how to find an *average*, or *mean*, see page 187.

Check It Out

Solve.

10 A trapezoid has a height of 4 meters. Its bases measure 5 meters and 8 meters. What is its area?

11 A trapezoid has a height of 8 centimeters. Its bases measure 7 centimeters and 10 centimeters. What is its area?

AREA

6·5

6·5 Exercises

1. Estimate the area of the shaded region.

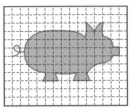

2. If each of the square units covered by the region measures 2 cm², estimate the area in centimeters.

Find the area of each rectangle given length, ℓ, and width, w.

3. ℓ = 5 ft, w = 4 ft
4. ℓ = 7.5 in., w = 6 in.

Find the area of each figure.

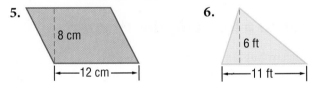

5.   8 cm |——12 cm——|

6.   6 ft |——11 ft——|

Find the area of each triangle, given base, b, and height, h.

7. b = 12 cm, h = 9 cm
8. b = 8 yd, h = 18 yd

9. Find the area of a trapezoid whose bases are 10 inches and 15 inches and whose height is 8 inches.

10. This is a design for a new pigpen. If the farmer builds it to have these measurements, how much area inside the pen will there be for the pigs?

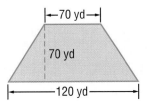

|←70 yd→|

70 yd

|————120 yd————|

6·6 Surface Area

The **surface area** of a solid is the total
area of its exterior surfaces. You can
think about surface area in terms of the
parts of a solid shape that you would
paint. Like area, surface area is expressed
in square units. To see why, "unfold" the rectangular prism.

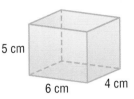

5 cm

6 cm   4 cm

Mathematicians call
this unfolded prism
a **net**. A net can be
folded to make a
three-dimensional
figure.

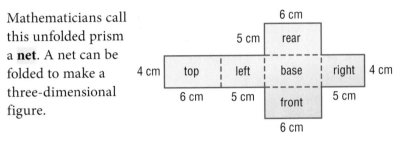

Surface Area of a Rectangular Prism

A rectangular prism has six rectangular faces. To find the
surface area of a rectangular prism, find the sum of the areas of
the six faces, or rectangles. Remember, opposite faces are equal.

EXAMPLE   Finding the Surface Area of a Rectangular Prism

Use the net to find the area of the rectangular prism above.

- Use the formula $A = \ell w$ to find the area of each face.
- Then add the six areas.
- Express the answer in square units.

Area	=	top + base	+	left + right	+	front + rear
	=	$2 \times (6 \times 4)$	+	$2 \times (5 \times 4)$	+	$2 \times (6 \times 5)$
	=	2×24	+	2×20	+	2×30
	=	48	+	40	+	60

$48 + 40 + 60 = 148$

The surface area of the rectangular prism is 148 cm².

Find the surface area of each shape.

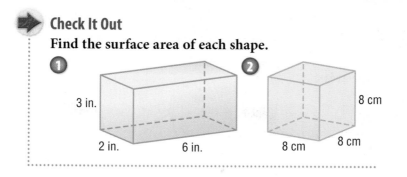

1 3 in. 2 in. 6 in.

2 8 cm 8 cm 8 cm

Surface Area of Other Solids

The unfolding technique can be used to find the surface area of any polyhedron. Look at the **triangular prism** and its net.

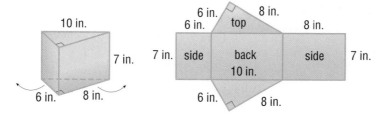

To find the surface area of this solid, use the area formulas for a rectangle ($A = \ell w$) and a triangle ($A = \frac{1}{2}bh$) to find the areas of the five faces. Then find the sum of the areas.

The area of the left side is 7 in. × 6 in. = 42 in².

The area of the back is 7 in. × 10 in. = 70 in².

The area of the right side is 7 in. × 8 in. = 56 in².

The area of the triangular top and bottom are each
$\frac{1}{2}$(6 in. × 8 in.) = 24 in².

The total area of the net is
42 in² + 56 in² + 70 in² + 24 in² × 2 = 216.

So, the surface area of the triangular prism is 216 in².

Below are two pyramids and their nets. For these polyhedrons, you again use the area formulas for a rectangle ($A = \ell w$) and a triangle ($A = \frac{1}{2}bh$) to find the areas of the faces. Then find the sum of the areas.

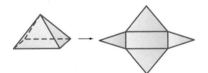

Rectangular pyramid

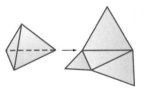

Tetrahedron (triangular pyramid)

Check It Out

Use the polyhedrons for Exercises 3 and 4.

3 Find the surface area of this rectangular prism.

4 mm

12 mm

6 mm

4 Unfold this triangular prism and find its surface area.

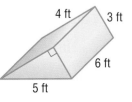

4 ft   3 ft

6 ft

5 ft

6·6 Exercises

1. What is an unfolded three-dimensional figure called?

Find the surface area of each shape.

2.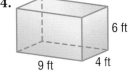
4 cm   3 cm
5 cm
7 cm

3. 10 mm
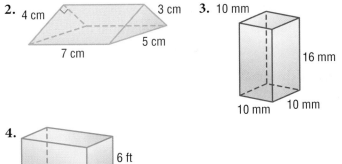
16 mm
10 mm   10 mm

4.
6 ft
9 ft   4 ft

5. A rectangular prism measures 5 cm × 3 cm × 7 cm. Find its surface area.

6. The surface area of a cube is 150 in². What is the length of an edge?
   A. 5 in.   **B.** 6 in.   **C.** 7 in.   **D.** 8 in.

7. What three-dimensional figure does the net at right represent?

8. What is the surface area of the figure in Exercise 7?

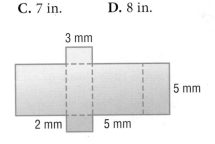
3 mm

5 mm

2 mm   5 mm

9. What is the surface area of a cube with sides that measure 3 meters?

10. Toshi is painting a box that is 4 feet long, 1 foot wide, and 1 foot high. He wants to put three coats of lacquer on it. He has a can of lacquer that will cover 200 square feet. Does he have enough lacquer for three coats?

6•7 Volume

What Is Volume?

Volume is the space inside a solid. One way to measure volume is to count the number of cubic units that would fill the space inside a solid.

The volume of this small cube is 1 cubic inch.

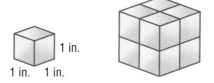

1 in.

1 in.   1 in.

The number of smaller cubes that it takes to fill the space inside the larger cube is 8, so the volume of the larger cube is 8 cubic inches.

You measure the volume of shapes in *cubic* units. For example, 1 cubic inch is written as 1 in^3, and 1 cubic meter is written as 1 m^3.

➡️ **Check It Out**

What is the volume of each shape if 1 cube = 1 in^3?

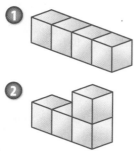

❶

❷

Volume of a Prism

The volume of a prism can be found by multiplying the *area* (pp. 281–286) of the base, *B*, and the height, *h*.

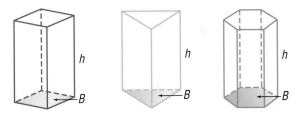

Volume = *Bh*

See *Formulas,* pages 58–59.

EXAMPLE **Finding the Volume of a Prism**

Find the volume of the rectangular prism. The base is 12 inches long and 10 inches wide; the height is 15 inches.

15 in.

12 in.   10 in.

base A = 12 in. × 10 in.   • Find the area of the base.

$\quad$ = 120 in^2

V = 120 in^2 × 15 in.   • Multiply the base and the height.

$\quad$ = 1,800 in^3 • Solve.

The volume of the prism is 1,800 in^3.

➡️ **Check It Out**

Find the volume of each prism.

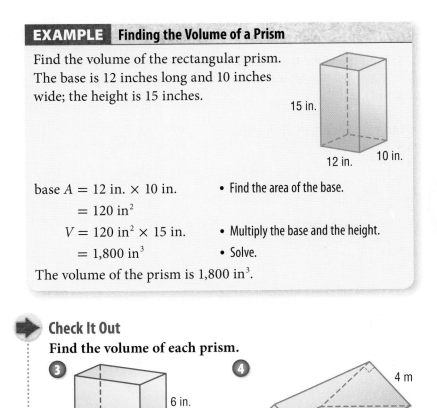

❸   6 in.

12 in.   3 in.

❹   4 m

5 m

9 m

6·7 Exercises

Use the rectangular prism to answer Exercises 1–4.

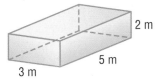

1. How many 1-meter cubes would it take to make one layer in the bottom of the prism?
2. How many layers of cubes would it take to fill the prism?
3. How many cubes would it take to fill the prism?
4. Each cube has a volume of 1 m³. What is the volume of the prism?

5. Find the volume of a rectangular prism with a base that measures 3 feet by 4 feet and with a height of 6 feet.
6. What is the formula for the volume of a prism?

Find the volume of each prism.

7.

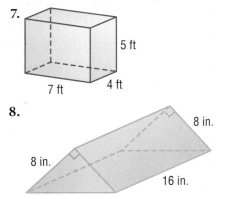

8.

6·8 Circles

Parts of a Circle

Of the many shapes you may encounter in geometry, circles are unique. They differ from other geometric shapes in several ways. For instance, while all circles are the same shape, polygons vary in shape. Circles have no sides, while polygons are named and classified by the number of sides they have. The *only* thing that makes one circle different from another is size.

A **circle** is a set of points equidistant from a given point. That point is the *center of the circle*. A circle is named by its center point.

Circle *P*

A **radius** is a **segment** that has one endpoint at the center and the other endpoint on the circle. In circle *P*, $\overline{PW}$ is a radius, and so is $\overline{PG}$.

A **diameter** is a line segment that passes through the center of the circle and has both endpoints on the circle. $\overline{GW}$ is a diameter of circle *P*. Notice that the length of the diameter $\overline{GW}$ is equal to the sum of $\overline{PW}$ and $\overline{PG}$. So, the diameter is twice the length of the radius. The diameter of circle *P* is 2(5) or 10 cm.

Check It Out
Find the radius or diameter.

1. Find the radius of a circle with diameter 12 cm.
2. Find the radius of a circle in which $d = y$.
3. Find the diameter of a circle with radius 10 in.
4. Find the diameter of a circle with radius 5.5 m.
5. Express the diameter of a circle whose radius is equal to *x*.

Circumference

The **circumference** of a circle is the distance around the circle. The *ratio* (p. 236) of every circle's circumference to its diameter is always the same. That ratio is a number that is close to 3.14. In other words, in every circle, the circumference is about 3.14 times the diameter. The symbol π, which is read as *pi*, is used to represent the ratio $\frac{C}{d}$.

$$\frac{C}{d} \approx 3.141592\ldots \text{Circumference} = \text{pi} \times \text{diameter, or } C = \pi d$$

Look at the illustration below. The circumference of the circle is about the same length as three diameters. This is true for any circle.

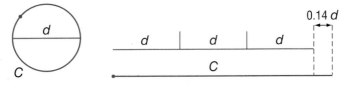

Because $d = 2r$, Circumference $= 2 \times \text{pi} \times \text{radius, or } C = 2\pi r$.

If you are using a calculator that has a π key, press that key to get an approximation for π to several more decimal places: π ≈ 3.141592. . . . For practical purposes, however, when you are finding the circumference of a circle, round π to 3.14, or simply leave the answer in terms of π.

EXAMPLE **Finding the Circumference of a Circle**

Find the circumference of a circle with a radius of 8 meters.

$d = 8 \times 2 = 16$ • Use the formula $C = \pi d$. Remember

$C = 16\pi$ to multiply the radius by 2 to get the diameter. Round the answer to the nearest tenth.

$C \approx 16 \times 3.14$ • The exact circumference is 16π m.

≈ 50.24

To the nearest tenth, the circumference is 50.2 meters.

You can find the diameter if you know the circumference. Divide both sides by π.

$$C = \pi d \qquad \frac{C}{\pi} = \frac{\pi d}{\pi} \qquad \frac{C}{\pi} = d$$

➡️ **Check It Out**

Find the radius, diameter, or circumference.

6 Find the circumference of a circle with a diameter of 8 millimeters. Give your answer in terms of π.

7 Find the circumference of a circle with a radius of 5 meters. Round to the nearest tenth.

8 Find the diameter to the nearest tenth of a circle with a circumference of 44 feet.

9 Find the radius of a circle with a circumference of 56.5 centimeters. Round your answer to the nearest whole number.

APPLICATION **Around the World**

Your blood vessels are a network of arteries and veins that carry oxygen to every part of the body and return blood with carbon dioxide to your lungs. Altogether, there are approximately 60,000 miles of blood vessels in the human body.

Just how far is 60,000 miles? The circumference of Earth is about 25,000 miles. If the blood vessels in one human body were placed end to end, approximately how many times would they wrap around the equator? See **HotSolutions** for the answer.

Area of a Circle

To find the area of a circle, you use the formula: Area = pi × radius squared, or $A = \pi r^2$. As with the area of polygons, the area of a circle is expressed in square units.

For a review of *area* and *square units,* see page 282.

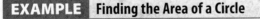

EXAMPLE **Finding the Area of a Circle**

Find the area of circle Q to the nearest whole number.

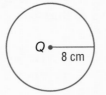

$A = \pi \times 8^2$ • Use the formula $A = \pi r^2$.

    $= 64\pi$ • Square the radius.

    ≈ 200.96 • Multiply by 3.14, or use the calculator

    $\approx 201 \text{ cm}^2$ key for π for a more exact answer.

The area of circle Q is about 201 cm².

If you are given the diameter instead of the radius, remember to divide the diameter by two.

Check It Out

Solve.

10 The diameter of a circle is 14 mm. Express the area of the circle in terms of π. Then multiply and round to the nearest tenth.

11 Use your calculator to find the area of a circle with a diameter of 16 ft. Use the calculator key for π or use $\pi \approx 3.14$, and round your answer to the nearest square foot.

6·8 Exercises

Find the diameter of each circle with the given radius.

1. 6 yd
2. 3.5 cm
3. 2.25 mm

Find the radius of each circle with the given diameter.

4. 20 ft
5. 13 m
6. 8.44 mm

Given the diameter or radius, find the circumference of the circle to the nearest tenth. Use 3.14 for π.

7. $d = 10$ in.
8. $d = 11.2$ m
9. $r = 3$ cm

The circumference of a circle is 88 cm. Find the following, to the nearest tenth.

10. the diameter
11. the radius

Find the area of each circle with the given radius or diameter. Round to the nearest whole number. Use 3.14 for π.

12. $d = 4.5$ ft
13. $r = 5$ in.
14. $d = 42$ m
15. $r = 16$ m

16. Which figure has the greater area: a circle with a diameter of 8 centimeters or a square that measures 7 centimeters on a side?

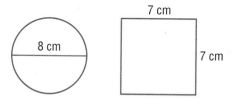

17. Sylvia has a circular glass-top table that measures 42 inches in diameter. When she dropped a jar on the table, the glass broke. Now she needs to order replacement glass by the square inch. To the nearest inch, how many square inches of glass will Sylvia need?

Geometry

What have you learned?

You can use the problems and the list of words that follow to see what you learned in this chapter. You can find out more about a particular problem or word by referring to the topic number (*for example*, Lesson 6·2).

Problem Set

Use this figure for Exercises 1–4. (Lesson 6·1)

1. Name an angle.
2. Name a ray.
3. What kind of angle is ∠*HIJ*?
4. Angle *HIK* measures 20°. What is the measure of ∠*HIJ*?

Use this figure for Exercises 5 and 6. (Lesson 6·2)

5. What kind of figure is polygon *ABCDE*?
6. Angle *A* measures 108°. What is the measure of ∠*B*?

For Exercises 7–9, choose the correct polyhedron. (Lesson 6·2)

7. pentagonal pyramid
8. square pyramid
9. pentagonal prism

10. What is the perimeter of a regular hexagon that measures 5 inches on a side? (Lesson 6·4)

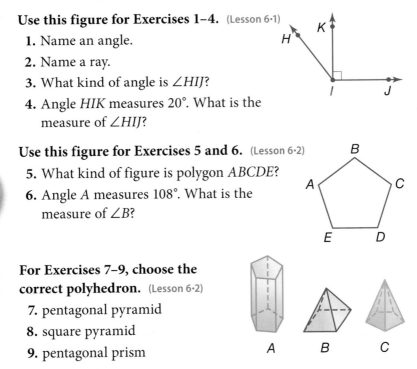

11. Find the area of a parallelogram with a base of 19 meters and a height of 6 meters. (Lesson 6·5)

12. Find the surface area of a rectangular prism whose length is 32 meters, width is 20 meters, and height is 5 meters. (Lesson 6·6)

13. Find the volume of a rectangular prism whose dimensions are 12.5 inches, 6.3 inches, and 2.7 inches. (Lesson 6·7)

Use circle _M_ to answer Exercises 14–16. (Lesson 6·8)

14. What is the diameter of circle _M_?

15. What is the area of circle _M_ to the nearest whole millimeter?

16. What is the circumference of circle _M_?

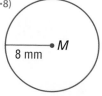

8 mm   • _M_

HotWords

Write definitions for the following words.

adjacent angles (Lesson 6·1)

angle (Lesson 6·1)

circumference (Lesson 6·6)

complementary angles (Lesson 6·1)

congruent angles (Lesson 6·1)

cube (Lesson 6·2)

degree (Lesson 6·1)

diagonal (Lesson 6·2)

diameter (Lesson 6·8)

face (Lesson 6·2)

net (Lesson 6·6)

opposite angle (Lesson 6·2)

parallelogram (Lesson 6·2)

perimeter (Lesson 6·4)

pi (Lesson 6·8)

point (Lesson 6·2)

polygon (Lesson 6·1)

polyhedron (Lesson 6·1)

prism (Lesson 6·2)

pyramid (Lesson 6·2)

quadrilateral (Lesson 6·2)

radius (Lesson 6·8)

ray (Lesson 6·1)

rectangular prism (Lesson 6·2)

reflection (Lesson 6·3)

regular polygon (Lesson 6·2)

right angle (Lesson 6·1)

rotation (Lesson 6·3)

supplementary angles (Lesson 6·1)

surface area (Lesson 6·5)

symmetry (Lesson 6·3)

transformation (Lesson 6·3)

translation (Lesson 6·3)

triangular prism (Lesson 6·6)

vertex (Lesson 6·1)

vertical angles (Lesson 6·1)

volume (Lesson 6·7)

HotTopic 7

Measurement

What do you know?

You can use the problems and the list of words that follow to see what you already know about this chapter. The answers to the problems are in **Hot**Solutions at the back of the book, and the definitions of the words are in **Hot**Words at the front of the book. You can find out more about a particular problem or word by referring to the topic number (*for example,* Lesson 7·2).

Problem Set

Write the correct metric system units for the following. (Lesson 7·1)

1. one hundredth of a liter

2. one thousand grams

Convert each of the following. (Lesson 7·2)

3. 100 mm = ____ cm

4. 15 yd = ____ ft

Find the perimeter of the rectangle below in the units indicated. (Lesson 7·2)

5. feet

6. inches

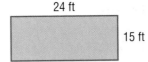
24 ft

15 ft

Find the area of the rectangle in the units indicated. (Lesson 7·3)

7. square feet

8. square inches

Convert the following measurements. (Lesson 7·3)

9. $2 \text{ m}^2 =$ ____ cm^2

10. $\frac{1}{2} \text{ ft}^2 =$ ____ in^2

11. $1 \text{ yd}^3 =$ ____ ft^3

12. $1 \text{ cm}^3 =$ ____ mm^3

13. 1,000 mL = ____ L

14. 2 gal = ____ qt

Imagine that you are going to summer camp and that you have packed your things in this trunk. The packed trunk weighs 96 pounds. (Lessons 7·3 and 7·4)

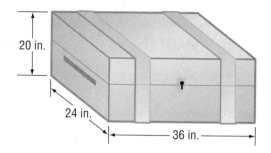

20 in.

24 in.

36 in.

15. What is the volume of the trunk in cubic feet?

16. The shipping company can bill you at either of two rates: $5.00 per cubic foot or $0.43 per pound. At which rate would you prefer to be billed? Why?

17. When you get to camp, you get a 3 feet × 3 feet × 3 feet space in which to store the trunk. Will it fit?

A team photograph that measures 3 inches × 5 inches needs to be enlarged so that it can be displayed in the school trophy case. (Lesson 7·6)

18. If the photo is enlarged to measure 6 inches × 10 inches, what would be the scale factor?

19. If the photo is reduced by a scale factor of $\frac{1}{2}$, what would be the measurements of the new photo?

HotWords

accuracy (Lesson 7·1)

area (Lesson 7·1)

capacity (Lesson 7·3)

customary system (Lesson 7·1)

distance (Lesson 7·2)

length (Lesson 7·2)

metric system (Lesson 7·1)

power (Lesson 7·1)

ratio (Lesson 7·6)

scale factor (Lesson 7·6)

side (Lesson 7·1)

similar figures (Lesson 7·6)

square (Lesson 7·1)

volume (Lesson 7·3)

7·1 Systems of Measurement

If you have ever watched the Olympic Games, you may have noticed that the distances are measured in meters or kilometers, and weights are measured in kilograms. That is because the most common system of measurement in the world is the **metric system**. In the United States, we typically use the **customary system** of measurement. It may be useful for you to make conversions from one unit of measurement to another within each system.

The Metric and Customary Systems

The metric system is based on **powers** of ten, such as 10,000 and 1,000. Converting within the metric system is simple because it is easy to multiply and divide by powers of ten.

Prefixes in the metric system have consistent meanings.

Prefix	Meaning	Example
milli-	one thousandth	1 *milli*liter = 0.001 liter
centi-	one hundredth	1 *centi*meter = 0.01 meter
kilo-	one thousand	1 *kilo*gram = 1,000 grams

Basic Measures			
	Metric		**Customary**
Distance	meter		inch, foot, yard, mile
Capacity	liter		cup, quart, gallon
Weight	gram		ounce, pound, ton

The customary system of measurement is not based on powers of ten. It is based on numbers like 12 and 16, which have many factors. This makes it easy to find, say, $\frac{2}{3}$ foot or $\frac{3}{4}$ pound. While the metric system uses decimals, you will frequently encounter fractions in the customary system.

Unfortunately, there are no convenient prefixes in the customary system, so you need to memorize the basic conversions: 16 oz = 1 lb; 36 in. = 1 yd; 4 qt = 1 gal; and so on.

Check It Out

Identify the measurement system.

1 Which system is based on multiples of 10?

2 Which system uses fractions?

Accuracy

Accuracy describes how close a measured value is to the actual (true) value. Every measurement is rounded to a certain place value. For example, you can measure to the nearest meter, tenth of a meter, and so on.

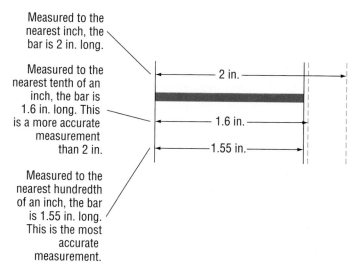

Measured to the nearest inch, the bar is 2 in. long.

Measured to the nearest tenth of an inch, the bar is 1.6 in. long. This is a more accurate measurement than 2 in.

Measured to the nearest hundredth of an inch, the bar is 1.55 in. long. This is the most accurate measurement.

2 in.

1.6 in.

1.55 in.

The more place values after the decimal that are used in a measurement, the closer the measurement is to the actual value. Measurements are always rounded up or down from the next possible place value.

Measurement	Possible Value
1.1 in.	between 1.05 in. and 1.14 in.
3 m	between 2.5 m and 3.4 m
6.25 ft	between 6.245 ft and 6.254 ft

The ability to measure with accuracy requires measuring tools which measure in the smallest possible increments, or smallest place value.

The length of each **side** of the **square** below is measured to the nearest tenth of a meter. But the actual length could be anywhere from 12.15 meters to 12.24 meters, because these are the numbers that round to 12.2.

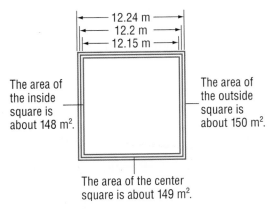

The area of the inside square is about 148 m².

The area of the outside square is about 150 m².

The area of the center square is about 149 m².

Since the side of the square could actually measure between 12.15 meters and 12.24 meters, the actual **area** may range from 148 square meters to 150 square meters.

Check It Out

Each side of a square measures 6.3 centimeters (to the nearest tenth).

③ The actual length of the side may range from ___ cm to ___ cm.

④ The actual area of the square may range from ___ cm² to ___ cm².

7·1 Exercises

What is the meaning of each metric system prefix?

1. milli-

2. centi-

3. kilo-

Write *customary* or *metric* to identify the system of measurement for the following.

4. ounces and pounds

5. meters and grams

6. feet and miles

7. Which measurement system uses fractions? Which uses decimals?

The measure of the side of the square is given to the nearest tenth. Express the actual area of the square as a range of measurements rounded to the nearest whole unit.

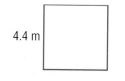

4.4 m

8. The actual area may range from ___ to ___.

10.2 in.

9. The actual areas may range from ___ to ___.

7·2 Length and Distance

About What Length?

When you get a feel for "about how long" or "around how far," it is easier to make estimations about **length** and **distance**. Here are some everyday items that will help you estimate metric and customary units.

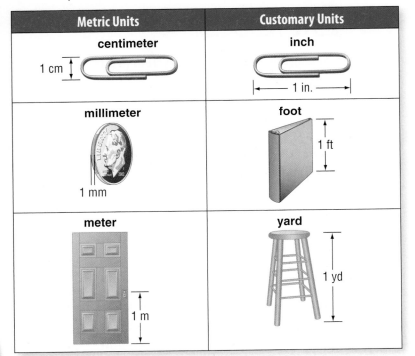

Metric Units	Customary Units
centimeter 1 cm	**inch** 1 in.
millimeter 1 mm	**foot** 1 ft
meter 1 m	**yard** 1 yd

➡️ **Check It Out**

Measure common items.

❶ Use a metric ruler or meterstick to measure common items. Name an item that is about a millimeter; about a centimeter; about a meter.

❷ Use a customary ruler or yardstick to measure common items. Name an item that is about an inch; about a foot; about a yard.

Metric and Customary Units of Length

When you calculate length and distance, you may use the metric system or the customary system. The commonly used metric measures for length and distance are millimeter (mm), centimeter (cm), meter (m), and kilometer (km). The customary system uses inch (in.), foot (ft), yard (yd), and mile (mi). Commonly used equivalents for both systems are shown below.

Metric Equivalents for Length			
1 km =	1,000 m =	100,000 cm =	1,000,000 mm
0.001 km =	1 m =	100 cm =	1,000 mm
	0.01 m =	1 cm =	10 mm
	0.001 m =	0.1 cm =	1 mm

Customary Equivalents for Length			
1 mi =	1,760 yd =	5,280 ft =	63,360 in.
$\frac{1}{1,760}$ yd =	1 yd =	3 ft =	36 in.
	$\frac{1}{3}$ yd =	1 ft =	12 in.
	$\frac{1}{36}$ yd =	$\frac{1}{12}$ ft =	1 in.

EXAMPLE   Changing Units Within a System

How many inches are in $\frac{1}{4}$ mile?

units you have

$\frac{1}{4}$ mi = ____ in.

1 mi = 63,360 in.

conversion factor

$\frac{1}{4} \times 63,360 = 15,840$

- Find the conversion statement from the unit you have to the unit you want.

- Because 63,360 in. = 1 mi, $\frac{1}{4} \times 63,360$ equals the number of inches in $\frac{1}{4}$ mi.

There are 15,840 inches in $\frac{1}{4}$ mile.

Check It Out

Convert each of the following.

3 72 in. to ft

4 50 mm to cm

APPLICATION From Boos to Cheers

It took 200 skyjacks two years and 2.5 million rivets to put together the Eiffel Tower. When it was completed in 1899, the art critics of Paris considered it a blight on the landscape. Today, it is one of the most familiar and beloved monuments in the world.

The tower's height, not counting its TV antennas, is 300 meters—that's about 300 yards or 3 football fields. Visitors can take elevators to the platforms or climb up the stairs: all **1,652** of them! On a clear day, the view can extend for 67 kilometers. How far does the view extend in meters? See **HotSolutions** for the answer.

7·2 Exercises

Convert each of the following.

1. 880 yd = ____ mi

2. $\frac{1}{4}$ mi = ____ ft

3. 9 ft = ____ in.

4. 1,500 m = ____ km

5. 17 cm = ____ m

6. 10 mm = ____ cm

What metric unit would you use to measure these objects?

7.

8.

What customary unit would you use to measure these objects?

9.

10.

7·3 Area, Volume, and Capacity

Area

Area is the measure, in square units, of a surface. The walls in your room are surfaces. The large surface of the United States takes up an area of 3,618,770 square miles. The area that the surface of a tire contacts on a wet road can make the difference between skidding or staying in control.

Area can be measured in metric units or customary units. Sometimes you may convert measurements within a measurement system. You can determine the conversions by using the basic measurements of length. Below is a chart that provides the most common conversions.

Metric Equivalents	Customary Equivalents
$100 \text{ mm}^2 = 1 \text{ cm}^2$	$144 \text{ in}^2 = 1 \text{ ft}^2$
$10,000 \text{ cm}^2 = 1 \text{ m}^2$	$9 \text{ ft}^2 = 1 \text{ yd}^2$
	$4,840 \text{ yd}^2 = 1 \text{ acre}$
	$640 \text{ acre} = 1 \text{ mi}^2$

To convert to a new unit, find the conversion statement for the units you have. Then multiply the units you have by the conversion factor for the new units. If the United States covers an area of about 3,800,000 mi^2, how many acres does it cover?

$1 \text{ mi}^2 = 640$ acres,

so, $3,800,000 \text{ mi}^2 \longrightarrow 3,800,000 \times 640 = 2,432,000,000$ acres.

 Check It Out

Convert the measurements.

1. Five square centimeters is equal to how many square millimeters?

2. Three square feet is equal to how many square inches?

Volume

Volume is expressed in cubic units. Here are some common relationships among units of volume.

Metric Equivalents	Customary Equivalents
$1{,}000 \text{ mm}^3 = 1 \text{ cm}^3$	$1{,}728 \text{ in}^3 = 1 \text{ ft}^3$
$1{,}000{,}000 \text{ cm}^3 = 1 \text{ m}^3$	$27 \text{ ft}^3 = 1 \text{ yd}^3$

EXAMPLE **Converting Volume Within a Measurement System**

Express the volume of the carton in cubic meters.

40 cm

50 cm

102 cm

$V = lwh$
$\quad = 120 \times 150 \times 40$
$\quad = 240{,}000 \text{ cm}^3$

$1{,}000{,}000 \text{ cm}^3 = 1 \text{ m}^3$

$240{,}000 \div 1{,}000{,}000 = 0.24 \text{m}^3$

The volume of the carton is 0.24 m³.

- Use a formula to find the volume (p. 292), using the units of the dimensions.
- Find the conversion factor.
- Multiply to convert to smaller units. Divide to convert to larger units.
- Include the unit of measurement in your answer.

Check It Out

Find the volume.

3 What is the volume of a box that measures 10 inches × 20 inches × 25 inches? Give your answer in cubic feet rounded to the nearest tenth.

4 What is the volume of a box that measures 4 centimeters on a side? Give your answer in cubic millimeters.

7·3

AREA, VOLUME, AND CAPACITY

Capacity

Capacity is closely related to volume. A block of wood has volume but no capacity to hold liquid. The capacity of a container is a measure of the volume of liquid it can hold.

Metric Equivalents		Customary Equivalents	
1 liter (L) =	1,000 milliliters (mL)	8 fl oz =	1 cup (c)
1 L =	1.057 qt	2 c =	1 pint (pt)
		2 pt =	1 quart (qt)
		4 qt =	1 gallon (gal)

Note the use of *fluid ounce* (fl oz) in the table. This is to distinguish it from *ounce* (oz) which is a unit of weight (16 oz = 1 lb). The fluid ounce is a unit of capacity (16 fl oz = 1 pint). There is a connection between ounce and fluid ounce, however. A pint of water weighs about a pound, so a fluid ounce of water weighs about an ounce. For most liquids used in cooking, *fluid ounce* and *ounce* are equivalent, and the "fl" is sometimes omitted (for example, "8 oz = 1 cup"). To be specific, use *ounce* to express weight only and *fluid ounce* to express capacity.

In the metric system, the basic unit of capacity is the *liter* (L). A liter has the capacity of about 1 quart.

$$1 \text{ L} = 1.057 \text{ qt}$$

Suppose that gasoline is priced at $1.17 per liter. How much is the price per gallon? There are 4 quarts in a gallon, so there are 4 × 1.057 or 4.228 liters in a gallon. So, a gallon of gasoline costs $1.17 × 4.228, or $4.947 per gallon.

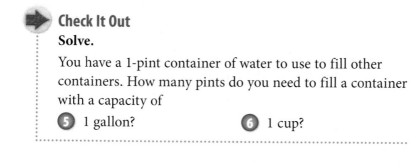

Check It Out

Solve.

You have a 1-pint container of water to use to fill other containers. How many pints do you need to fill a container with a capacity of

5 1 gallon?

6 1 cup?

7.3 Exercises

Identify whether the unit of measure is used to express distance, area, or volume.

1. square centimeter
2. mile
3. cubic meter
4. kilometer

For Exercises 5–7, express the volume of this box in the units indicated.

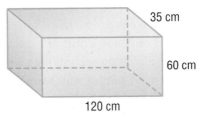

35 cm

60 cm

120 cm

5. cubic centimeters
6. cubic millimeters
7. cubic meters

Convert each of the following measurements.

8. 1 L = ____ mL   9. 4 qt = ____ gal
10. 1,057 qt = ____ L   11. 500 mL = ____ L
12. 2 gal = ____ c   13. 1 qt = ____ fl oz
14. 3 qt = ____ gal   15. 8 c = ____ qt
16. 5 c = ____ qt   17. 16 fl oz = ____ c
18. 5,000 mL = ____ L

19. When Sujey said, "It holds about 2 liters," was she talking about a bathtub, a bottle of cola, or a paper cup?

20. When Pei said, "It holds about 250 gallons," was he pointing to an oil tank, a milk container, or an in-ground swimming pool?

7·4 Mass and Weight

Mass and *weight* are not equivalent. Mass is the amount of matter in a substance. Weight is the pull of gravity on the substance.

Your mass is the same on the Moon as it is here on Earth. But, if you weigh 100 pounds on Earth, you weigh about $16\frac{2}{3}$ pounds on the Moon. That is because the gravitational pull of the Moon is only $\frac{1}{6}$ that of Earth.

Metric Equivalents	Customary Equivalents
1 kg = 1,000 g = 1,000,000 mg	1 T = 2,000 lb = 32,000 oz
0.001 kg = 1 g = 1,000 mg	0.0005 T = 1 lb = 16 oz
0.000001 kg = 0.001 g = 1 mg	0.0625 lb = 1 oz
1 lb ≈ 0.4536 kg	
1 kg ≈ 2.205 lb	

To convert from one unit of mass or weight to another, first find the conversion statement for the units you have in the equivalents chart. Then multiply or divide the number of units you have by the conversion factor for the new units.

EXAMPLE Converting Units of Mass

If you have 64 ounces of peanut butter, how many pounds of peanut butter do you have?

$1 \text{ oz} = 0.0625 \text{ lb}$ • Find the pound equivalent for 1 ounce.

$64 \text{ oz} = (64 \times 0.0625) \text{ lb}$ • Multiply.

$\phantom{64 \text{ oz}} = 4 \text{ lb}$

You have 4 pounds of peanut butter.

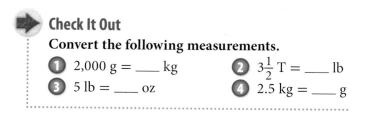

Check It Out

Convert the following measurements.

1 2,000 g = ____ kg

2 $3\frac{1}{2}$ T = ____ lb

3 5 lb = ____ oz

4 2.5 kg = ____ g

7·4 Exercises

Convert the following measurements.

1. 7 kg = _____ mg
2. 500 mg = _____ g
3. 1,000 lb = _____ T
4. 8 oz = _____ lb
5. 200,000 mg = _____ kg
6. 2 T = _____ lb
7. 2,000 mg = _____ kg
8. 6 kg = _____ lb
9. 15 lb = _____ kg
10. 100 lb = _____ oz
11. $\frac{1}{2}$ lb = _____ oz
12. 5 g = _____ mg
13. 4,000 g = _____ kg
14. 48 oz = _____ lb
15. 2.5 g = _____ mg
16. 2 T = _____ oz

17. The weight of a baby born in Democratic Republic of the Congo was recorded as 2.7 kilograms. If the baby had been born in the United States, what weight would have been recorded in pounds?

18. Nirupa's scale shows her weight in both pounds and kilograms. She weighed 95 pounds one morning. How many kilograms did she weigh?

19. Coffee is on sale at Slonim's for $7.99 per pound. The same brand is advertised at Harrow's for $7.99 per 0.5 kilograms. Which store has the better buy?

20. A baker needs 15 pounds of flour to make bread. How many 2-kilogram bags of flour will she need?

7·5 Time

Time measures the interval between two or more events. You can measure time with a very short unit—a second—or a very long unit—a millennium—and with many units in between.

Time Equivalents	
60 seconds (sec) = 1 minute (min)	365 d = 1 yr
60 min = 1 hour (hr)	10 yr = 1 decade
24 hr = 1 day (d)	100 yr = 1 century
7 d = 1 week (wk)	1,000 yr = 1 millennium
12 months (mo) = 1 year (yr)	(*Millennia* means "more than one millennium.")

- 1,000,000 seconds before 12:00 A.M., January 1, 2000 is 10:13:20 A.M., December 20, 1999.
- 1,000,000 hours before 12:00 A.M., January 1, 2000 is 8:00 A.M., December 8, 1885.

Like other kinds of measurement, you can convert one unit of time to another, using the information in the table above.

EXAMPLE Converting Units of Time

If Hulleah is 13 years old, what is her age in months?

1 yr = 12 mo	• Find the month equivalent for one year.
13 yr = 13 × 12 mo	• Multiply.
= 156 mo	

So, Hulleah is 156 months old.

Check It Out

Convert the units of time.

1. How many months old will you be on your twenty-first birthday?

2. What year will it be 5,000 days from January 1, 2010?

7·5 Exercises

Convert each of the following units.

1. 2 d = ____ hr
2. 90 min = ____ hr
3. 1 yr = ____ d
4. 200 yr = ____ centuries
5. 5 min = ____ sec
6. 30 centuries = ____ millennia

7. How many days are in three 365-day years?

8. How many minutes are in a day?

9. How many hours are in a week?

10. How many years old will you be when you have lived for 6,939 days?

APPLICATION | **The World's Largest Reptile**

Would it surprise you to learn that the world's largest reptile is a turtle? The leatherback turtle can weigh as much as 2,000 pounds. By comparison, an adult male crocodile weighs about 1,000 pounds.

The leatherback has existed in its current form for over 20 million years, but this prehistoric giant is now endangered. If after 20 million years of existence the leatherback was to become extinct, how many times longer than *Homo sapiens* will it have existed? Assume *Homo sapiens* have been around 4,000 millennia. See **HotSolutions** for the answer.

7·6 Size and Scale

Similar Figures

Similar figures are figures that have exactly the same shape but are not necessarily the same size. The corresponding side lengths of similar figures are in proportion.

EXAMPLE Deciding Whether Two Figures Are Similar

Are these two rectangles similar?

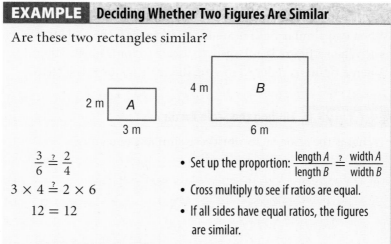

$$\frac{3}{6} \stackrel{?}{=} \frac{2}{4}$$

$$3 \times 4 \stackrel{?}{=} 2 \times 6$$

$$12 = 12$$

• Set up the proportion: $\frac{\text{length } A}{\text{length } B} \stackrel{?}{=} \frac{\text{width } A}{\text{width } B}$

• Cross multiply to see if ratios are equal.

• If all sides have equal ratios, the figures are similar.

So, the rectangles are similar.

Check It Out

Find similar figures.

1 Which figures are similar to the shaded figure?

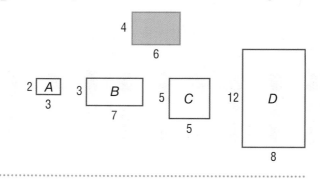

Scale Factors

A **scale factor** is the ratio of two corresponding sides of two similar figures.

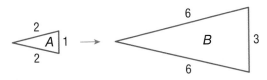

$\triangle A$ is similar to $\triangle B$. $\triangle B$ is 3 times larger than $\triangle A$. The scale factor is 3.

When two similar figures are identical in size, the scale factor is 1. When a figure is enlarged, the scale factor is greater than 1. When a figure is reduced in size, the scale factor is less than 1.

EXAMPLE Finding the Scale Factor

What is the scale factor for these similar pentagons?

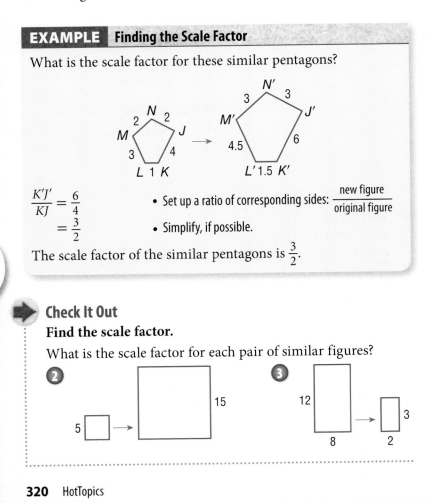

$\dfrac{K'J'}{KJ} = \dfrac{6}{4}$
$= \dfrac{3}{2}$

- Set up a ratio of corresponding sides: $\dfrac{\text{new figure}}{\text{original figure}}$
- Simplify, if possible.

The scale factor of the similar pentagons is $\dfrac{3}{2}$.

Check It Out

Find the scale factor.

What is the scale factor for each pair of similar figures?

2

5 → 15

3

12 → 3

8   2

7·6 Exercises

Give the scale factor for each pair of similar figures.

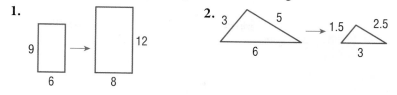

1.

9, 6, 12, 8

2.

3, 5, 6, 1.5, 2.5, 3

Given the scale factors, find the missing dimensions for the sides of the figures.

3. The scale factor is 3.

4. The scale factor is $\frac{1}{2}$.

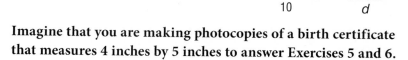

4, 5, 3, b, c, a

8, 10, d, e

Imagine that you are making photocopies of a birth certificate that measures 4 inches by 5 inches to answer Exercises 5 and 6.

5. Enlarge the birth certificate by a scale factor of 2. What are the dimensions of the enlargement?

6. Reduce the birth certificate by a scale factor of $\frac{1}{2}$. What are the dimensions of the reduction?

A map is drawn to the scale 1 cm = 2 km.

7. If the distance between the school and the post office measures 5 centimeters on the map, what is the actual distance between them?

8. If the actual distance between the library and the theater is 5 kilometers, how far apart do they appear on the map?

9. The highway that runs from one end of the city to the opposite end measures 3.5 centimeters on the map. What is the actual distance of this section of the highway?

Measurement

What have you learned? You can use the problems and the list of words that follow to see what you learned in this chapter. You can find out more about a particular problem or word by referring to the topic number (*for example,* Lesson 7·2).

Problem Set

Write the correct metric system unit for each of the following. (Lesson 7·1)

1. one thousandth of a meter
2. one hundredth of a liter
3. one thousand grams

Convert each of the following measurements. (Lesson 7·2)

4. 350 mm = ____ m

5. 0.07 m = ____ mm

6. 6 in. = ____ yd

Give the perimeter of the rectangle below in the units indicated. (Lesson 7·2)

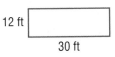

12 ft

30 ft

7. in feet

8. in inches

Give the area of the rectangle in the units indicated. (Lesson 7·3)

9. square feet

10. square inches

Convert the following measurements of area, volume, and capacity. (Lesson 7·3)

11. 10 m^2 = ____ cm^2

12. 4 ft^2 = ____ in^2

13. 3 yd^3 = ____ ft^3

14. 4 cm^3 = ____ mm^3

15. 3,000 mL = ____ L

16. 6 gal = ____ qt

A family photograph that measures 4 inches by 6 inches needs to be enlarged so that it can be displayed in a new frame. (Lesson 7·6)

17. If the photo is enlarged to measure 12 inches by 18 inches, what is the scale factor?

18. If the photo is enlarged by a scale factor of 2, what are the measurements of the new photo?

Imagine that you are going to summer camp and that you have packed your things in the trunk shown below. The packed trunk weighs 106 lb. (Lessons 7·3 and 7·4)

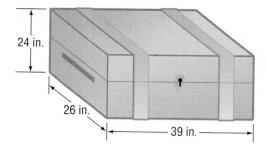

24 in.

26 in.

39 in.

19. What is the volume of the trunk in inches?

20. The shipping company can bill you at either of two rates; $4.00 per cubic foot or $0.53 per pound. At which rate would you prefer to be billed? Why?

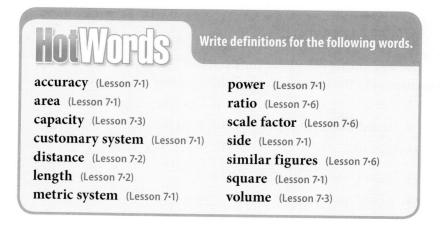

HotWords

Write definitions for the following words.

accuracy (Lesson 7·1)

area (Lesson 7·1)

capacity (Lesson 7·3)

customary system (Lesson 7·1)

distance (Lesson 7·2)

length (Lesson 7·2)

metric system (Lesson 7·1)

power (Lesson 7·1)

ratio (Lesson 7·6)

scale factor (Lesson 7·6)

side (Lesson 7·1)

similar figures (Lesson 7·6)

square (Lesson 7·1)

volume (Lesson 7·3)

HotTopic 8

Tools

What do you know?	You can use the problems and the list of words that follow to see what you already know about this chapter. The answers to the problems are in **HotSolutions** at the back of the book, and the definitions of the words are in **HotWords** at the front of the book. You can find out more about a particular problem or word by referring to the topic number (*for example,* Lesson 8·2).

Problem Set

Use your calculator to complete Exercises 1–6. (Lesson 8·1)

1. $40 + 7 \times 5 + 4^2$

2. 300% of 450

Round answers to the nearest tenth.

3. $8 + 3.75 \times 5^2 + 15$

4. $62 + (-30) \div 0.5 - 12.25$

5. Find the perimeter of rectangle *ABCD*.

6. Find the area of rectangle *ABCD*.

Rectangle *ABCD* with vertices *A* (top left), *B* (top right), *C* (bottom right), *D* (bottom left); right side labeled 4 cm, bottom labeled 8.5 cm.

Use a scientific calculator for Exercises 7–12. Round decimal answers to the nearest hundredth. (Lesson 8·2)

7. 5.5^3

8. Find the reciprocal of 8.

9. Find the square of 12.4.

10. Find the square root of 4.5.

11. $(2 \times 10^3) \times (9 \times 10^2)$

12. $7.5 \times (6 \times 3.75)$

13. What is the measure of $\angle VRT$? (Lesson 8·3)

14. What is the measure of $\angle SRV$? (Lesson 8·3)

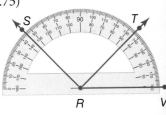

15. Does $\overrightarrow{RT}$ divide $\angle SRV$ into two equal angles? (Lesson 8·3)

Refer to the spreadsheet for Exercises 16–18. (Lesson 8·4)

16. Name the cell holding 1,000.

17. A formula for cell B2 is B1 × 10. Name another formula for cell B2.

18. Cell D1 contains the number 1,000 and no formula. After using the command *fill down,* what number will be in cell D3?

HotWords

angle (Lesson 8·3)	**perimeter** (Lesson 8·4)
cell (Lesson 8·4)	**pi** (Lesson 8·1)
circle (Lesson 8·1)	**power** (Lesson 8·2)
column (Lesson 8·4)	**radius** (Lesson 8·1)
cube (Lesson 8·2)	**ray** (Lesson 8·3)
cube root (Lesson 8·2)	**reciprocal** (Lesson 8·2)
decimal (Lesson 8·2)	**root** (Lesson 8·2)
distance (Lesson 8·3)	**row** (Lesson 8·4)
factorial (Lesson 8·2)	**spreadsheet** (Lesson 8·4)
formula (Lesson 8·4)	**square** (Lesson 8·2)
horizontal (Lesson 8·4)	**square root** (Lesson 8·1)
negative number (Lesson 8·1)	**vertex** (Lesson 8·3)
parentheses (Lesson 8·2)	**vertical** (Lesson 8·4)
percent (Lesson 8·1)	

8·1 Four-Function Calculator

People use calculators to make mathematical tasks easier. You may have seen your parents using a calculator to balance a checkbook. But a calculator is not always the fastest way to do a mathematical task. If your answer does not need to be exact, it might be faster to estimate. Sometimes you can do the problem in your head quickly, or using pencil and paper might be a better method. Calculators are particularly helpful for problems with many numbers or with numbers that have many digits.

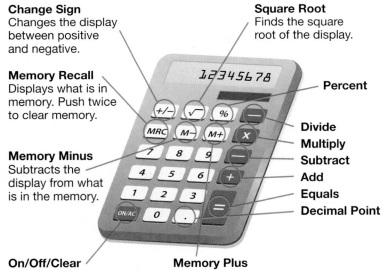

Change Sign
Changes the display between positive and negative.

Square Root
Finds the square root of the display.

Memory Recall
Displays what is in memory. Push twice to clear memory.

Percent

Divide

Multiply

Memory Minus
Subtracts the display from what is in the memory.

Subtract

Add

Equals

Decimal Point

On/Off/Clear
Turns the calculator on or off. Clears the display.

Memory Plus
Adds the display to what is in the memory.

A calculator only gives the answer to the problem that you enter. Always have an estimate of the answer you expect. Then compare the calculator answer to your estimate to be sure that you entered the problem correctly.

Basic Operations

Adding, subtracting, multiplying, and dividing are fairly straightforward calculator operations.

Operation	Problem	Calculator Keys	Display
Addition	10.5 + 39	10.5 [+] 39 [=]	49.5
Subtraction	40 − 51	40 [−] 51 [=]	−11.
Multiplication	20.5 × 4	20.5 [×] 4 [=]	82.
Division	12 ÷ 40	12 [÷] 40 [=]	0.3.

Negative Numbers

To enter a **negative number** into your calculator, press [+/−] after you enter the number.

Problem	Calculator Keys	Display
−15 + 10	15 [+/−] [+] 10 [=]	−5.
50 − (−32)	50 [−] 32 [+/−] [=]	82.
−9 × 8	9 [+/−] [×] 8 [=]	−72.
−20 ÷ (−4)	20 [+/−] [÷] 4 [+/−] [=]	5.

Check It Out

Find each answer on a calculator.

1 11.6 + 4.2

2 45.4 − 13.9

3 20 × (−1.5)

4 −24 ÷ 0.5

Memory

For complex or multi-step problems, you can use the memory function on your calculator. You operate the memory with three special keys. The way many calculators operate is shown on page 328. If yours does not work this way, check the instructions that came with your calculator.

Key	Function
MRC	Press once to display (recall) what is in memory. Press twice to clear memory.
M+	Adds display to what is in memory.
M−	Subtracts display from what is in memory.

When the calculator memory contains something other than zero, the display will show $\boxed{^M}$ along with whatever number the display currently shows. What you do on your calculator does not change memory unless you use the special memory keys.

You could use the following keystrokes to calculate $10 + 55 + 26 \times 2 + 60 - 4^2$ on your calculator:

Keystrokes	Display
MRC MRC C	0.
4 × 4 M−	M 16.
26 × 2 M+	M 52.
10 + 55 M+	M 65.
60 M+	M 60.
MRC	161.

Your answer is 161. Notice the use of the *order of operations* (p. 76).

➡️ **Check It Out**

Use the memory function on your calculator to find each answer.

5️⃣ $5 \times 10 - 18 \times 3 + 8^2$

6️⃣ $-40 + 5^2 - (-14) \times 6$

7️⃣ $6^3 \times 4 + (-18) \times 20 + (-50)$

8️⃣ $20^2 + 30 \times (-2) - (-60)$

Special Keys

Some calculators have keys with special functions.

Key	Function
$\boxed{\sqrt{x}}$	Finds the **square root** of the display.
$\boxed{\%}$	Changes display to the decimal expression of a **percent**.
$\boxed{\pi}$	Automatically enters **pi** to as many places as your calculator holds.

The $\boxed{\%}$ and $\boxed{\pi}$ keys save you time by saving keystrokes.
The $\boxed{\sqrt{}}$ key finds square roots more precisely, something
difficult to do with paper and pencil. See how these keys work
in the examples below.

Problem: $10 + \sqrt{144}$
Keystrokes: $10 \boxed{+} 144 \boxed{\sqrt{}} \boxed{=}$
Final display: 22.

If you try to find the square root of a negative number, your
calculator will display an error message. For example, the
keystrokes $64 \boxed{+/-} \boxed{\sqrt{}}$ result in the display $\boxed{\text{E} \qquad 5.}$. There
is no square root of -64, because no number multiplied by itself
can give a negative number.

Problem: Find 40% of 50.
Keystrokes: $50 \boxed{\times} 40 \boxed{\%}$
Final display: 20.

The $\boxed{\%}$ key changes a percent to its decimal form. If you know
how to convert percents to decimals, you probably will not use
the $\boxed{\%}$ key often.

Problem: Find the area of a circle with radius 2.
(Use the formula $A = \pi r^2$.)
Keystrokes: $\boxed{\pi} \boxed{\times} 2 \boxed{\times} 2 \boxed{=}$
Final display: 12.57

If your calculator does not have a $\boxed{\pi}$ key, you can use 3.14
or 3.1416 as an approximation for π.

Answer the following.

9 Without using the calculator, tell what displays if you enter: 12 [M+] 4 [×] 2 [+] [MRC] [=].

10 Use memory functions to calculate $160 - 8^2 \times (-6)$.

11 Find the square root of 196.

12 Find 25% of 450.

APPLICATION **Calculator Alphabet**

You probably think calculators are only useful for doing arithmetic. But you can also send "secret" messages with them—if you know the calculator alphabet. Try this. Enter the number 0.7734 in your calculator and turn it upside down. What word appears in the display? See **HotSolutions** for the answer.

Each of the calculator's numerical keys can be used to display a letter.

- **8** to display a **B**
- **6** to display a **g**
- **4** to display an **h**
- **7** to display an **L**
- **5** to display an **S**
- **3** to display an **E**

- **9** to display a **G** or **b**
- **1** to display an **I**
- **0** to display an **O**
- **2** to display a **Z**

See how many words you can display with your calculator. Remember, you must turn the calculator upside down to read the words, so enter the numbers in reverse order.

FOUR-FUNCTION CALCULATOR

8·1

8·1 Exercises

Find the value of each expression, using your calculator.

1. $15.6 + 22.4$
2. $45.61 - 20.8$
3. $-16.5 - 5.6$
4. $10 \times 45 \times 30$
5. $-5 + 60 \times (-9)$
6. $50 - 12 \times 20$
7. $\sqrt{81} - 16$
8. $-10 + \sqrt{225}$
9. $12 \div 20 + 11$
10. $12 \div (-20)$
11. 20% of 350
12. 120% of 200
13. $216 - \sqrt{484}$
14. $\sqrt{324} \div 2.5 + 8.15$

Use a calculator to answer Exercises 15–21.

15. Find the area if $x = 2.5$ cm.
16. Find the perimeter if $x = 4.2$ cm.

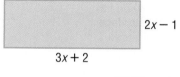

17. Find the area if $a = 1.5$ in.
18. Find the circumference if $a = 3.1$ in.

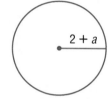

19. Find the area of $\triangle RQP$.
20. Find the circumference of circle Q.
21. Find the area of circle Q.

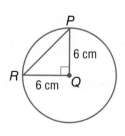

8•2 Scientific Calculator

Every mathematician and scientist has a scientific calculator to solve complex equations quickly and accurately. Scientific calculators vary widely; some have just a few functions and others have many functions. Some calculators can even be programmed to perform functions you choose. The calculator below shows functions you might find on a scientific calculator.

2nd
Press to get the 2nd function for any key. 2nd functions are listed above each key, typically on the calculator face.

Square Root
Finds the square root of the display.

Display

π
Automatically enters π.

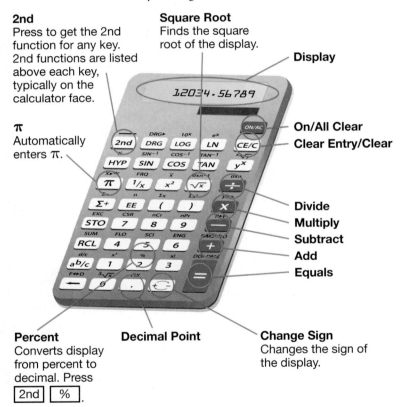

On/All Clear

Clear Entry/Clear

Divide

Multiply

Subtract

Add

Equals

Percent
Converts display from percent to decimal. Press 2nd %.

Decimal Point

Change Sign
Changes the sign of the display.

Frequently Used Functions

Because each scientific calculator is set up differently, your calculator may not work exactly as below. These keystrokes work with the calculator illustrated on page 332. Use the reference book or card that came with your calculator to perform similar functions. See the index to find more information about the mathematics here.

Function	Problem	Keystrokes
Cube Root $\sqrt[3]{x}$ Finds the cube root of the display.	$\sqrt[3]{64}$	64 [2nd] [$\sqrt[3]{x}$] [4.]
Cube x^3 Finds the cube of the display.	5^3	5 [2nd] [x^3] [125.]
Factorial $x!$ Finds the factorial of the display.	$5!$	5 [2nd] [$x!$] [120.]
Fix [FIX] Rounds the display to the number of places you enter.	Round 3.729 to the hundredths place.	3.729 [2nd] [FIX] 2 [3.73]
Parentheses [( ] [ )] Groups calculations.	$8 \times (7 + 2)$	8 [×] [( ] [7] [+] 2 [ )] [=] [72.]
Powers y^x Finds the x power of the display.	12^4	12 [y^x] 4 [=] [20736.]
Powers of ten [10^x] Raises ten to the power displayed.	10^3	3 [2nd] [10^x] [1000.]

Function	Problem	Keystrokes
Reciprocal $\boxed{1/x}$ Finds the reciprocal of the display.	Find the reciprocal of 10.	$10 \boxed{1/x} \boxed{ 0.1}$
Roots $\boxed{\sqrt[x]{y}}$ Finds the x root of the display.	$\sqrt[4]{1296}$	$1296 \boxed{2\text{nd}} \boxed{\sqrt[x]{y}} 4 \boxed{=}$ $\boxed{ 6.}$
Square $\boxed{x^2}$ Finds the square of the display.	9^2	$9 \boxed{x^2} \boxed{ 81.}$

Check It Out

Use your calculator to find the following.

1 $\sqrt[3]{91.125}$ **2** 7^3

3 $7!$ **4** 9^4

Use your calculator to find the following to the nearest thousandth.

5 $6 \times (21 - 3) \div (2 \times 5)$ **6** the reciprocal of 2

7 19^2 **8** $\sqrt[3]{512} \times 4^4 + \sqrt{400}$

8·2 Exercises

Use a scientific calculator to find the following. Round your answer to the nearest hundredth, if necessary.

1. 18^2 **2.** 9^3 **3.** 12^3 **4.** 2.5^2

5. 3π **6.** $\dfrac{30}{\pi}$ **7.** $\dfrac{1}{5}$ **8.** $\dfrac{2}{\pi}$

9. $(10 + 4.1)^2 + 4$

10. $12 - (20 \div 2.5)$

11. $\sqrt[4]{1296}$

12. reciprocal of 40

8·3 Geometry Tools

Ruler

If you need to measure the dimensions of an object, or if you need to measure reasonably short **distances**, use a ruler.

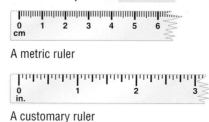

A metric ruler

A customary ruler

To get an accurate measure, be sure that one end of the item being measured lines up with zero on your ruler.

The pencil below is measured first to the nearest tenth of a centimeter and then to the nearest eighth of an inch.

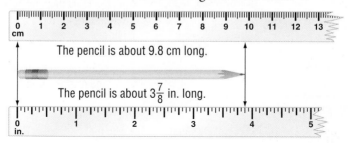

The pencil is about 9.8 cm long.

The pencil is about $3\frac{7}{8}$ in. long.

> ### Check It Out
> Use your ruler to measure each line segment to the nearest tenth of a centimeter or to the nearest eighth of an inch.
>
> 1 ──────────
>
> 2 ──────────────
>
> 3 ────────────────────
>
> 4 ───────

Protractor

Measure **angles** with a *protractor*. There are many different protractors. The key is to find the point on each protractor to which you align the **vertex** of the angle.

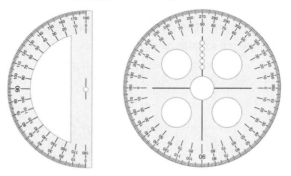

EXAMPLE **Measuring Angles with a Protractor**

Use the protractor to find *m∠CDE* and *m∠FDC*.

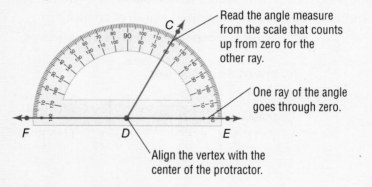

Read the angle measure from the scale that counts up from zero for the other ray.

One ray of the angle goes through zero.

Align the vertex with the center of the protractor.

So, *m∠CDE* = 60° and *m∠FDC* = 120°.

To draw an angle with a given measure, draw one **ray** first and position the center of the protractor at the endpoint. Then mark a dot at the desired measure. Finally, draw a ray from the vertex to the dot to form the angle.

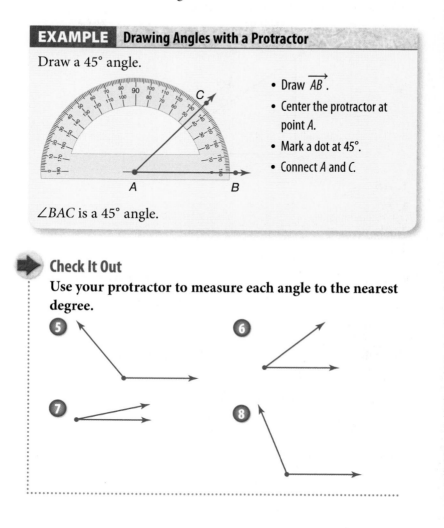

EXAMPLE Drawing Angles with a Protractor

Draw a 45° angle.

- Draw $\overrightarrow{AB}$.
- Center the protractor at point *A*.
- Mark a dot at 45°.
- Connect *A* and *C*.

∠*BAC* is a 45° angle.

Check It Out

Use your protractor to measure each angle to the nearest degree.

5

6

7

8

8·3 Exercises

Using a ruler, measure the length of each side of △ABC. Give your answer in inches or centimeters, rounded to the nearest $\frac{1}{8}$ in. or $\frac{1}{10}$ cm.

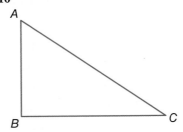

1. *AB*

2. *BC*

3. *AC*

Use a protractor to measure each angle in △ABC.

4. ∠*C* 5. ∠*B* 6. ∠*A*

7. What is the sum of the measures of the interior angles of any triangle?

Fill in the blank with the name of the correct tool.

8. A _____ is used to measure distance.

9. A _____ is use to measure angles.

Measure the following angles.

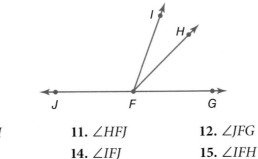

10. ∠*GFH* 11. ∠*HFJ* 12. ∠*JFG*

13. ∠*HFI* 14. ∠*IFJ* 15. ∠*IFH*

Measure the following angles.

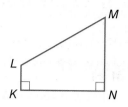

16. $\angle NML$

17. $\angle MLK$

18. $\angle KNM$

19. $\angle LKN$

20. Use a protractor to copy $\angle MLK$.

Using a ruler and a protractor, copy the figures below.

21.

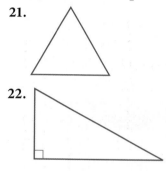

22.

8·4 Spreadsheets

What Is a Spreadsheet?

People have used **spreadsheets** as tools to keep track of information, such as finances, for a long time. Spreadsheets were paper-and-pencil math tools before becoming computerized. You may be familiar with computer spreadsheet programs.

A spreadsheet is a computer tool where information is arranged into **cells** within a grid and calculations are performed within the cells. When one cell is changed, all other cells that depend on it automatically change.

Spreadsheets are organized into **rows** and **columns**. Rows are **horizontal** and are numbered. Columns are **vertical** and are named by capital letters. A cell is named by its row and column.

File	Edit			
Sample.xls			🔲 🗗 ❎	
◇	**A**	**B**	**C**	**D**
1	1	3	1	
2	2	6	4	
3	3	9	9	
4	4	12	16	
5	5	15	25	
	Sheet 1	Sheet 2	Sheet 3	

The cell A3 is in Column A, Row 3. In this spreadsheet, there is a 3 in cell A3.

Check It Out

In the spreadsheet above, what number appears in each cell?

1 A5   2 B2   3 C2

Spreadsheet Formulas

A cell can contain a number or a word, or it may contain needed information to generate a number. A **formula** generates a number dependent on the entries in other cells in the spreadsheet. The way the formulas are written depends on the particular spreadsheet computer software you are using. You typically enter a formula and the value generated shows in the cell; the formula entered is behind the cell.

EXAMPLE Creating a Spreadsheet Formula

	A	B	C	D
1	Item	Price	Qty	Total
2	sweater	$30	2	$60
3	pants	$18	2	
4	shirt	$10	4	
5				
6				

←Express the value of the cell in relationship to other cells.

$$Total = Price \times Qty$$
$$D2 = B2 \times C2$$

If you change the value of a cell and a formula depends on that value, the result of the formula will change.

In the spreadsheet above, if you entered 3 sweaters instead of 2 (C2 = 3), the total column would automatically change to $90.

Check It Out

Use the spreadsheet above. If the total is always figured the same way, write the formula for:

4 D3

5 D4

6 What is the price of one shirt?

7 What is the total spent on sweaters?

Fill Down and Fill Right

Now that you know the basics, look at some ways to make spreadsheets do even more of the work for you. *Fill down* and *fill right* are two spreadsheet commands that can save a lot of time and effort.

To use *fill down,* select a portion of a column. *Fill down* will take the top cell that has been selected and copy it into the lower cells. If the top cell in the selected range contains a number, such as 5, *fill down* will generate a column of 5s.

If the top cell of the selected range contains a formula, the *fill down* feature will automatically adjust the formula as you go from cell to cell.

The selected cells are highlighted.

File	Edit	
Sample.xls		
◇	A	B
1	100	
2	= A1 + 10	
3		
4		
5		
6		
7	Sheet 1 / Sheet 2	

The spreadsheet fills the column and adjusts the formula.

File	Edit	
Sa	Fill down	
	Fill right	
◇	A	B
1	100	
2	= A1 + 10	
3	= A2 + 10	
4	= A3 + 10	
5	= A4 + 10	
6		
7	Sheet 1 / Sheet 2	

These are the values that actually appear.

File	Edit		
Sa Fill down **s**			▢ ◲ ☒
Fill right			
◇	A	B	⌃
1	100∎		
2	110		
3	120		≣
4	130		
5	140		
6			
⊲⊳ ◀ ▶ ▶| \ Sheet 1 / Sheet 2 /			⌄
⟨	III		⟩

Fill right works in a similar manner, except it goes across, copying the leftmost cell of the selected range in a row.

Row 1 is selected.

File	Edit					
Sa Fill down **s**					▢ ◲ ☒	
Fill right						
◇	A	B	C	D	E	⌃
1	100∎					
2	= A1 + 10					
3	= A2 + 10					≣
4	= A3 + 10					
5	= A4 + 10					
6						
⊲⊳ ◀ ▶ ▶| \ Sheet 1 / Sheet 2 /						⌄
⟨	III					⟩

The 100 fills to the right.

File	Edit					
Sa Fill down **s**					▢ ◲ ☒	
Fill right						
◇	A	B	C	D	E	⌃
1	100∎	100	100	100	100	
2	= A1 + 10					
3	= A2 + 10					≣
4	= A3 + 10					
5	= A4 + 10					
6						
⊲⊳ ◀ ▶ ▶| \ Sheet 1 / Sheet 2 /						⌄
⟨	III					⟩

If you select A1 to E1 and fill right, you will get all 100s.

If you select A2 to E2 and fill right, you will "copy" the formula A1 + 10 as shown.

File	Edit				
					□ ▣ ✕

Fill down
Fill right

◇	A	B	C	D	E	⌃
1	100	100	100	100	100	
2	= A1 + 10					≡
3						
4						
5						
6						

|◀ ◀ ▶ ▶| \ Sheet 1 / Sheet 2 / ⌄
< III >

Row 2 is selected.

File	Edit				
					□ ▣ ✕

Fill down
Fill right

◇	A	B	C	D	E	⌃
1	100	100	100	100	100	
2	= A1 + 10	= B1 + 10	= C1 + 10	= D1 + 10	= E1 + 10	
3	= A2 + 10					≡
4	= A3 + 10					
5	= A4 + 10					
6						

|◀ ◀ ▶ ▶| \ Sheet 1 / Sheet 2 / ⌄
< III >

The spreadsheet completes the calculations and fills the cells.

Check It Out

Use the spreadsheet above.

8 Select B1 to B5 and fill down. What number will be in B3?

9 Select A3 to C3 and fill right. What formula will be in C3? what number?

10 Select A4 to E4 and fill right. If D3 = 120, what formula will be in D4? what number?

11 Select E2 to E6 and fill down. If E5 = 140, what formula will be in E6? what number?

SPREADSHEETS

8·4

Spreadsheet Graphs

You can graph from a spreadsheet. As an example, use a spreadsheet that compares the **perimeter** of a square to the length of its side.

◇	A	B	C	D	E
1	side	perimeter			
2	1	4			
3	2	8			
4	3	12			
5	4	16			
6	5	20			
7	6	24			
8	7	28			
9	8	32			
10	9	36			
11	10	40			

File Edit — Sample.xls — Sheet 1 / Sheet 2

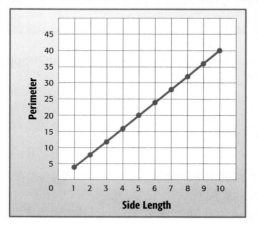

Most spreadsheets have a function that displays tables as graphs. See your spreadsheet reference for more information.

Use the spreadsheet and graph from page 345.

12 What cells give the point (1, 4)?

13 What cells give the point (5, 20)?

14 What point is shown by cells A9, B9?

15 What point is shown by cells A11, B11?

8·4 Exercises

File	Edit				

Sample Fill down / Fill right ☐ ☐ ☒

◇	A	B	C	D	▲
1	1	2	80	100	
2	2	4	70	200	
3	3	6	80	300	▥
4	4		90		
5					
6					▼

|◀ ◀ ▶ ▶|\ Sheet 1 / Sheet 2 /

◀ ||| ▶

For the spreadsheet shown above, what number appears in each of the following cells?

1. B2   **2.** A3   **3.** C1

In which cell does each number appear?

4. 200   **5.** 3   **6.** 90

7. If the formula behind cell C2 is C1 + 10, what formula is behind cell C3?

8. What formula might be behind cell B2?

9. Say cells D5 and D6 were filled by using fill down from D3. D3 has the formula D2 + 100. What would the values of D5 and D6 be?

10. The formula behind cell B2 is A2 × 2. What formula might be behind cell B3?

Use the spreadsheet below to answer Exercises 11–15.

	A	B	C
	File Edit		
1	1	5	10
2	= A1 + 2	= B1 − 1	= C1 × 2
3			
4			
5			

Fill down
Fill right

Sheet 1 / Sheet 2

11. If you select A2 to A4 and fill down, what formula will appear in A4?

12. If you select C2 to C5 and fill down, what numbers will appear in C2 to C5?

13. If you select A2 to C2 and fill right, what will appear in B2? *Hint:* Fill will cover over any existing numbers or formulas.

14. If you select A1 to D1 and fill right, what will appear in D1?

15. If you select B2 to B5 and fill down, what will appear in B4?

Tools

What have you learned?

You can use the problems and the list of words that follow to see what you learned in this chapter. You can find out more about a particular problem or word by referring to the topic number (for example, Lesson 8·2).

Problem Set

Use your calculator to answer Exercises 1–4. Round answers to the nearest tenth, if necessary. (Lesson 8·1)

1. $65 + 12 \times 3 + 6^2$

2. 250% of 740

3. $17 - 5.6 \times 8^2 + 24$

4. $122 - (-45) \div 0.75 - 9.65$

For Exercises 5–10, use a scientific calculator. Round decimal answers to the nearest hundredth. (Lesson 8·2)

5. 3.6^4

6. Find the reciprocal of 7.1.

7. Find the square of 7.5.

8. Find the square root of 7.5.

9. $(6 \times 10^5) \times (9 \times 10^4)$

10. $2.6 \times (13 \times 5.75)$

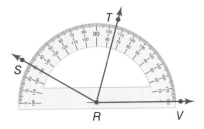

11. What is the measure of $\angle TRV$? (Lesson 8·3)

12. What is the measure of $\angle VRS$? (Lesson 8·3)

13. What is the measure of $\angle TRS$? (Lesson 8·3)

14. Does $\overrightarrow{RT}$ divide $\angle SRV$ into two equal angles? (Lesson 8·3)

15. What is the measure of an angle that bisects a right angle? (Lesson 8·3)

For Exercises 16–18, refer to the spreadsheet below. (Lesson 8·4)

File	Edit			
	Fill down			
Sa	Fill right			

◇	A	B	C	D
1	1	1	50	1
2	7	3	100	
3	13	9	150	

Sheet 1 / Sheet 2 /

16. Name the cell with entry 3.

17. A formula for cell C3 is = C2 + 50. Name another formula for cell C3.

18. Cell D1 contains the number 1 and no formula. After using the command *fill down*, what number will be in cell D4?

HotWords

angle (Lesson 8·3)

cell (Lesson 8·4)

circle (Lesson 8·1)

column (Lesson 8·4)

cube (Lesson 8·2)

cube root (Lesson 8·2)

decimal (Lesson 8·2)

distance (Lesson 8·3)

factorial (Lesson 8·2)

formula (Lesson 8·4)

horizontal (Lesson 8·4)

negative number (Lesson 8·1)

parentheses (Lesson 8·2)

percent (Lesson 8·1)

perimeter (Lesson 8·4)

pi (Lesson 8·1)

power (Lesson 8·2)

radius (Lesson 8·1)

ray (Lesson 8·3)

reciprocal (Lesson 8·2)

root (Lesson 8·2)

row (Lesson 8·4)

spreadsheet (Lesson 8·4)

square (Lesson 8·2)

square root (Lesson 8·1)

vertex (Lesson 8·3)

vertical (Lesson 8·4)

Hot Solutions

HotSolutions

Chapter ❶ Numbers and Computation

1. 30,000   **2.** 30,000,000
3. $(2 \times 10,000) + (4 \times 1,000) + (3 \times 100)$
   $+ (7 \times 10) + (8 \times 1)$
4. 566,418; 496,418; 56,418; 5,618
5. 52,564,760; 52,565,000; 53,000,000   **6.** 0
7. 15   **8.** 400   **9.** 1,600   **10.** $(4 + 6) \times 5 = 50$
11. $(10 + 14) \div (3 + 3) = 4$ **12.** no   **13.** no
14. no   **15.** yes   **16.** 3×11 **17.** $3 \times 5 \times 7$
18. $2 \times 2 \times 3 \times 3 \times 5$

19. 15   **20.** 7   **21.** 6   **22.** 15   **23.** 24   **24.** 80
25. 6   **26.** -13 **27.** 15   **28.** -25 **29.** 6   **30.** -1
31. -18 **32.** 0   **33.** 28   **34.** -4 **35.** 7   **36.** 36
37. -30 **38.** -60 **39.** It will be a negative integer.
40. It will be a positive integer.

1•1 Place Value of Whole Numbers

1. 30   **2.** 3,000,000   **3.** forty million, three hundred
   six thousand, two hundred   **4.** fourteen trillion,
   thirty billion, five hundred million

5. $(8 \times 10,000) + (3 \times 1,000) + (4 \times 10) + (6 \times 1)$
6. $(3 \times 100,000) + (2 \times 100) + (8 \times 10) + (5 \times 1)$

7. <   **8.** >   **9.** 6,520; 52,617; 56,302; 526,000
10. 32,400   **11.** 560,000   **12.** 2,000,000   **13.** 400,000

1•2 Properties

1. yes **2.** no   **3.** no   **4.** yes

1•3 Order of Operations

1•4 Factors and Multiples

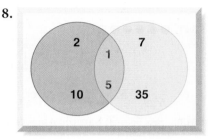
1•5 Integer Operations

Chapter ❷ Fractions, Decimals, and Percents

p. 94   **1.** \$45.75   **2.** 92%   **3.** C   **4.** $4\frac{1}{6}$ **5.** $1\frac{9}{20}$

6. $3\frac{5}{6}$ **7.** $4\frac{37}{63}$ **8.** $\frac{5}{16}$ **9.** $\frac{9}{65}$ **10.** hundredths

11. $(4 \times 1) + (6 \times 0.1) + (3 \times 0.001)$ **12.** 0.247

p. 95   **13.** 1.065; 1.605; 1.655; 16.5   **14.** 17.916   **15.** 13.223

16. 101.781   **17.** 25%   **18.** 5.3   **19.** 68%   **20.** 50%

21. 6%   **22.** 56%   **23.** 0.34   **24.** 1.25

2•1 Fractions and Equivalent Fractions

p. 97   **1.** $\frac{2}{3}$ **2.** $\frac{4}{9}$

3. Sample answer: ●●●●●○○○;

p. 98   **4.** Sample answer: $\frac{2}{6}, \frac{3}{9}$ **5.** Sample answer: $\frac{1}{2}, \frac{2}{4}$

6. Sample answer: $\frac{6}{10}, \frac{9}{15}$ **7.** Sample answer: $\frac{2}{2}, \frac{3}{3}, \frac{4}{4}$

p. 99   **8.** =   **9.** =   **10.** ≠

p. 101   **11.** $\frac{1}{5}$ **12.** $\frac{1}{3}$ **13.** $\frac{9}{10}$

p. 102   **Musical Fractions** $\frac{1}{32}, \frac{1}{64}$;

p. 104   **14.** $4\frac{4}{5}$ **15.** $1\frac{4}{9}$ **16.** $2\frac{3}{4}$ **17.** $4\frac{5}{6}$

18. $\frac{17}{10}$ **19.** $\frac{41}{8}$ **20.** $\frac{33}{5}$ **21.** $\frac{52}{7}$

2•2 Comparing and Ordering Fractions

p. 107   **1.** <   **2.** <   **3.** >   **4.** =   **5.** >   **6.** >   **7.** >

p. 108   **8.** $\frac{2}{4}; \frac{5}{8}; \frac{4}{5}$ **9.** $\frac{7}{12}; \frac{2}{3}; \frac{3}{4}$ **10.** $\frac{5}{8}; \frac{2}{3}; \frac{5}{6}$ **11.** 28   **12.** 126

2•3 Addition and Subtraction of Fractions

p. 110 1. 2 2. $\frac{6}{25}$ 3. $\frac{5}{23}$ 4. $\frac{3}{8}$

p. 112 5. $1\frac{1}{4}$ 6. $\frac{1}{6}$ 7. $\frac{7}{10}$ 8. $\frac{7}{12}$ 9. $9\frac{5}{6}$ 10. $34\frac{5}{8}$

11. 61 12. $26\frac{6}{7}$

p. 113 13. $6\frac{3}{8}$ 14. $60\frac{1}{6}$ 15. $59\frac{3}{8}$

p. 114 16. $6\frac{3}{8}$ 17. $16\frac{1}{6}$ 18. $17\frac{29}{30}$ 19. $8\frac{7}{10}$

p. 115 20. $3\frac{1}{2}$ or 3 21. 7 22. 9 23. $7\frac{1}{2}$ or 8

2•4 Multiplication and Division of Fractions

p. 118 1. $\frac{5}{18}$ 2. $\frac{15}{28}$ 3. $\frac{2}{9}$ 4. $\frac{33}{100}$ 5. $\frac{1}{6}$ 6. $\frac{3}{7}$ 7. $\frac{16}{45}$ 8. $\frac{2}{3}$

p. 119 9. $\frac{8}{3}$ 10. $\frac{1}{5}$ 11. $\frac{2}{9}$

p. 120 12. 8 13. $9\frac{19}{48}$ 14. $72\frac{11}{24}$ 15. $1\frac{1}{4}$ 16. $1\frac{3}{7}$ 17. $6\frac{2}{9}$

p. 121 18. $1\frac{1}{2}$ 19. $5\frac{1}{3}$ 20. $\frac{3}{4}$

2•5 Naming and Ordering Decimals

p. 124 1. 0.9 2. 0.55 3. 7.18 4. 5.03

5. $(0 \times 1) + (6 \times 0.1) + (3 \times 0.01) + (4 \times 0.001)$

6. $(3 \times 1) + (2 \times 0.1) + (2 \times 0.01) + (1 \times 0.001)$

7. $(0 \times 1) + (7 \times 0.01) + (7 \times 0.001)$

p. 126 8. five ones; five and six hundred thirty-three thousandths

9. five thousandths; forty-five thousandths

10. seven thousandths; six and seventy-four ten-thousandths

11. one hundred thousandth; two hundred seventy-one hundred-thousandths 12. < 13. < 14. >

p. 127 15. 4.0146, 4.1406, 40.146 16. 8, 8.073, 8.373, 83.037

17. 0.52112, 0.522, 0.5512, 0.552 18. 1.66 19. 226.95

20. 7.40 21. 8.59

2·6 Decimal Operations

p. 129 **1.** 88.88 **2.** 61.916 **3.** 6.13 **4.** 46.283

p. 130 **5.** 12 **6.** 4 **7.** 14 **8.** 13

p. 131 **9.** 4.704 **10.** 114.1244

p. 132 **11.** 0.001683 **12.** 0.048455 **13.** 210 **14.** 400

p. 133 **15.** 5 **16.** 4.68 **17.** 50.4 **18.** 46

p. 134 **19.** 0.73 **20.** 0.26 **21.** 0.60

2·7 Meaning of Percent

p. 136 **1.** 32% shaded; 68% not shaded
2. 44% shaded; 56% not shaded
3. 15% shaded; 85% not shaded

p. 137 **4.** 23 **5.** 40 **6.** 60 **7.** 27

p. 138 **8.** Sample answer: $1.45 **9.** Sample answer: $4
10. Sample answer: $19

2·8 Using and Finding Percents

p. 140 **1.** 19.25 **2.** 564 **3.** 12.1 **4.** 25.56

p. 141 **5.** 25% **6.** 15% **7.** 5% **8.** 150%

p. 142 **9.** 108 **10.** 20 **11.** 23.33 **12.** 925

p. 143 **13.** 36.4 **14.** 93.13 **15.** 11.16 **16.** 196

p. 144 **17.** Sample answer: 100 **18.** Sample answer: 2
19. Sample answer: 15 **20.** Sample answer: 30

p. 145 **Honesty Pays** 20%

2·9 Fraction, Decimal, and Percent Relationships

p. 148 **1.** 55% **2.** 40% **3.** 75% **4.** 43% **5.** $\frac{4}{25}$ **6.** $\frac{1}{25}$

7. $\frac{19}{50}$ **8.** $\frac{18}{25}$

p. 149 **9.** $\frac{49}{200}$ **10.** $\frac{67}{400}$ **11.** $\frac{969}{800}$

Chapter ❸ Powers and Roots

3•1 Powers and Exponents

3•2 Square Roots

Chapter ❹ Data, Statistics, and Probability

p. 170   **1.** no   **2.** unbiased   **3.** bar graph   **4.** Wednesday
    5. You cannot tell from this graph.

p. 171   **6.** $160   **7.** $138.20; $155   **8.** 26.8   **9.** 76
    10. $\frac{11}{20}$ **11.** $\frac{15}{260}$

4·1 Collecting Data

p. 172   **1.** students signed up for after-school sports; 60
    2. wolves on Isle Royale; 15

p. 173   **3.** Sample answer: Draw names out of a bag.
    4. No; it is limited to people who are in that fitness
    center, so they may like it best.

p. 174   **5.** It calls table tennis boring.
    6. It does not imply that you are more adventurous
    if you like the sport.
    7. Do you donate money to charity?

p. 175   **8.** 3   **9.** yes
    10. Have a party because most students said yes.

4·2 Displaying Data

p. 178   **1.** 18

    2.

Number of Sponsors	2	3	4	5	6	7	8	9	10
Number of Students	3	0	3	5	2	0	2	3	4

p. 180   **3.** nonfiction and videos
    4. Sample answer: The library has slightly more
    nonfiction than fiction in its collection.
    5. The video collection is exactly one third the size
    of the fiction collection.

p. 181 **6.** 5   **7.** 2

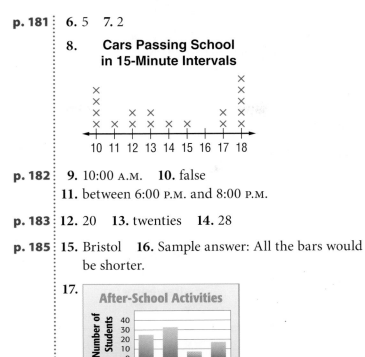

8. **Cars Passing School in 15-Minute Intervals**

p. 182 **9.** 10:00 A.M.   **10.** false

11. between 6:00 P.M. and 8:00 P.M.

p. 183 **12.** 20   **13.** twenties   **14.** 28

p. 185 **15.** Bristol   **16.** Sample answer: All the bars would be shorter.

17.

Graphic Impressions One might think the size of the pictures represents the *size* of the animals; the bar graph more accurately portrays the data.

4·3 Statistics

p. 188 **1.** 24.75   **2.** 87   **3.** 228   **4.** $37

p. 189 **5.** 18   **6.** 25   **7.** 96   **8.** 157.5 lb

p. 190 **9.** 53   **10.** 96 and 98   **11.** 14   **12.** no mode

p. 192 **13.** 18   **14.** 2

15. 20; It made the original mean a lesser value.

p. 193 **16.** 613   **17.** 75   **18.** 32°

How Mighty Is the Mississippi? 3,039.3 mi; 2,659 mi; 1,845 mi

4·4 Probability

p. 196
1. $\frac{3}{6}$ or $\frac{1}{2}$ 2. $\frac{8}{12}$ or $\frac{2}{3}$

3. $\frac{1}{4}$, 0.25, 1:4, 25%   4. $\frac{2}{5}$, 0.4, 2:5, 40%

p. 198

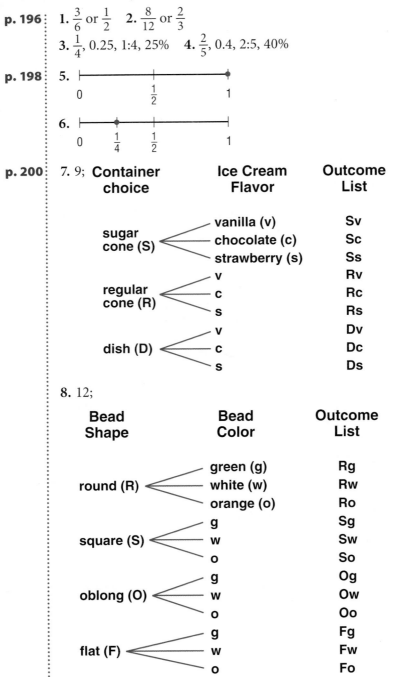

5.

6.

p. 200
7. 9;

Container choice	Ice Cream Flavor	Outcome List
sugar cone (S)	vanilla (v)	Sv
	chocolate (c)	Sc
	strawberry (s)	Ss
regular cone (R)	v	Rv
	c	Rc
	s	Rs
dish (D)	v	Dv
	c	Dc
	s	Ds

8. 12;

Bead Shape	Bead Color	Outcome List
round (R)	green (g)	Rg
	white (w)	Rw
	orange (o)	Ro
square (S)	g	Sg
	w	Sw
	o	So
oblong (O)	g	Og
	w	Ow
	o	Oo
flat (F)	g	Fg
	w	Fw
	o	Fo

9. 8;

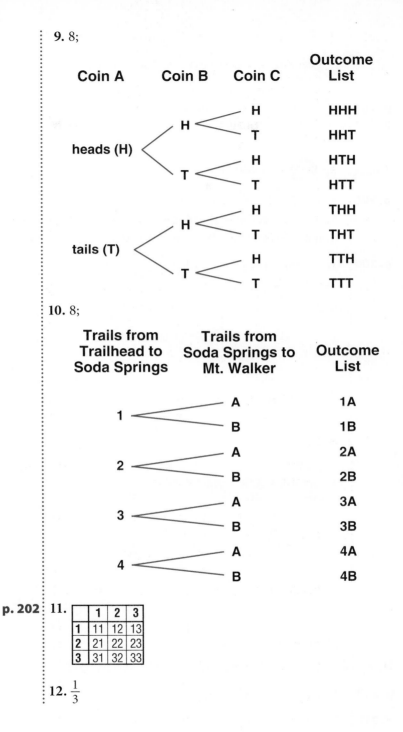

Coin A	Coin B	Coin C	Outcome List
heads (H)	H	H	HHH
		T	HHT
	T	H	HTH
		T	HTT
tails (T)	H	H	THH
		T	THT
	T	H	TTH
		T	TTT

10. 8;

Trails from Trailhead to Soda Springs	Trails from Soda Springs to Mt. Walker	Outcome List
1	A	1A
	B	1B
2	A	2A
	B	2B
3	A	3A
	B	3B
4	A	4A
	B	4B

p. 202 **11.**

	1	2	3
1	11	12	13
2	21	22	23
3	31	32	33

12. $\frac{1}{3}$

p. 203 **13.** $\frac{3}{40}$ **14.** $\frac{11}{20}$ **15.** Sample answer: My coin landed
heads-up 27 times. My neighbor's landed heads-up
24 times.

p. 204 **16.** $\frac{1}{3}$ **17.** 1 **18.** $\frac{1}{2}$ **19.** $\frac{5}{12}$

Chapter **5** Algebra

p. 208 **1.** $x + 5$ **2.** $4n$ **3.** $(x + 3) - 2$
4. $3x + 7$ **5.** $7n - 10$ **6.** 36 **7.** $5 + 3n = 22$
8. false, false, true **9.** 15 mi **10.** 10 boys

p. 209 **11.** 7.5 cm **12.** $x < 3$ **13.** $x \geq 4$ **14.** $n > 2$
15. $n \leq -2$

16–21.

22.

Input (x)	Output (x + 4)
1	5
2	6
3	7

5·1 Writing Expressions and Equations

p. 210 **1.** 2 **2.** 1 **3.** 3 **4.** 1

p. 211 **5.** $5 + x$ **6.** $n + 4$ **7.** $y + 8$ **8.** $n + 2$

p. 212 **9.** $8 - x$ **10.** $n - 3$ **11.** $y - 6$ **12.** $n - 4$

p. 213 **13.** $4x$ **14.** $8n$ **15.** $0.25y$ **16.** $9n$

p. 214 **17.** $\frac{x}{5}$ **18.** $\frac{8}{n}$ **19.** $\frac{20}{y}$ **20.** $\frac{a}{4}$

5·2 Simplifying Expressions

p. 216 **1.** no **2.** yes **3.** no **4.** yes

p. 217 **5.** $7 + 2x$ **6.** $6n$ **7.** $4y + 5$ **8.** $8 \cdot 3$

. . . 3, 2, 1 Blast Off Answers will vary depending on height. To match a flea's feat, a 5-foot-tall child would have to jump 800 feet.

p. 218 **9.** $4 + (5 + 8)$ **10.** $2 \cdot (3 \cdot 5)$ **11.** $5x + (4y + 3)$
12. $(6 \cdot 9)n$ **13.** $6(100 - 1) = 594$
14. $3(100 + 6) = 318$ **15.** $4(200 - 2) = 792$
16. $5(200 + 10 + 1) = 1{,}055$

p. 219 **17.** $6x + 2$ **18.** $12n - 18$

p. 220 **19.** $11x$ **20.** $6y$ **21.** $8n$ **22.** $10a$

p. 221 **23.** $8 + y$ **24.** $37 + n$ **25.** $20x$ **26.** $36n$

5·3 Evaluating Expressions and Formulas

p. 223 **1.** 11 **2.** 11 **3.** 27 **4.** 8

p. 224 **5.** 32 cm **6.** 18 ft **7.** 30 mi **8.** 1,200 km **9.** 220 mi
10. 6 ft

5·4 Equations

p. 227 **1.** $5x - 9 = 6$ **2.** $n + 6 = 10$ **3.** $4y - 1 = 7$

p. 228 **4.** true, false, false **5.** false, true, false
6. false, false, true **7.** false, false, true

p. 229 **8.** 3 **9.** 3 **10.** 15 **11.** $x + 3 = 15$ **12.** $x - 3 = 9$
13. $3x = 36$ **14.** $\frac{x}{3} = 4$

p. 231 **15.** 6 **16.** 10 **17.** 35 **18.** 8

p. 232 **19.** 5   **20.** 24   **21.** 4   **22.** 300

p. 234 **23.** 24, 36, 48   **24.** $f = x \div 2$

5·5 Ratio and Proportion

p. 236 **1.** $\frac{4}{8} = \frac{1}{2}$ **2.** $\frac{8}{12} = \frac{2}{3}$ **3.** $\frac{12}{4} = \frac{3}{1} = 3$

p. 237 **4.** \$.10 per min   **5.** 18 points per game
6. 3 cm per day   **7.** 100 ft per min

p. 238 **8.** yes   **9.** no   **10.** yes   **11.** yes

p. 239 **12.** 2.3 gal   **13.** \$105   **14.** \$13

5·6 Inequalities

p. 242 **1.**

2.

3.

4.

5. $x > 5$ **6.** $y > 9$

5·7 Graphing on the Coordinate Plane

p. 244 **1.** x-axis   **2.** Quadrant II   **3.** Quadrant IV   **4.** y-axis

p. 245 **5.** (3, 1)   **6.** (2, −4)   **7.** (−4, 0)   **8.** (0, 3)

p. 246 9–12.

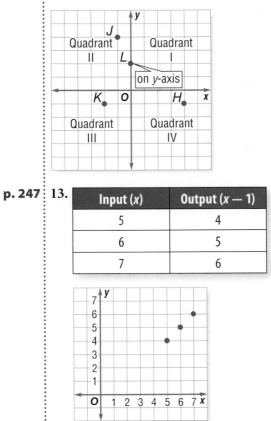

p. 247 13.

Input (x)	Output ($x - 1$)
5	4
6	5
7	6

Chapter ⑥ Geometry

p. 252 1. Sample answers: $\angle ABC$ or $\angle CBA$; $\angle ABD$ or $\angle DBA$; $\angle DBC$ or $\angle CBD$ 2. Possible answers: $\overrightarrow{BA}$, $\overrightarrow{BD}$, $\overrightarrow{BC}$
3. 45°   4. parallelogram   5. 360°   6. 16 ft   7. 84 cm^3
8. 256 in^2

p. 253 9. B   10. A   11. C   12. 60 cm^3 13. 6 in.
14. 28.3 in^2

6·1 Naming and Classifying Angles and Triangles

6·2 Polygons and Polyhedrons

6·3 Symmetry and Transformations

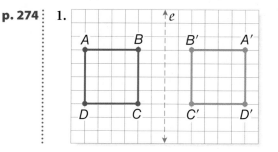

6•4 Perimeter

6•5 Area

6•6 Surface Area

6•7 Volume

6•8 Circles

Chapter **7** Measurement

p. 302 **1.** centiliter **2.** kilogram **3.** 10 **4.** 45 **5.** 78
6. 936 **7.** 360 **8.** 51,840 **9.** 20,000 **10.** 72
11. 27 **12.** 1,000 **13.** 1 **14.** 8

p. 303 **15.** 10 ft^3

16. $0.43 per pound; the per pound rate would cost
$8.72 less than the cubic foot rate.

17. yes **18.** 2 **19.** $1\frac{1}{2}$ in. $\times$ $1\frac{1}{2}$ in.

7·1 Systems of Measurement

p. 305 **1.** metric **2.** customary

p. 307 **3.** 6.25, 6.34 **4.** 39, 40

7·2 Length and Distance

p. 308 **1.** Sample answers: a pencil tip, a pencil eraser, a tennis
racquet
2. Sample answers: a nickel, a shoe box, a baseball bat

p. 309 **3.** 6 ft **4.** 5 cm

p. 310 **From Boos to Cheers** 6,700

7·3 Area, Volume, and Capacity

p. 311 **1.** 500 **2.** 432

p. 312 **3.** 2.9 ft^3 **4.** 64,000 mm^3

p. 313 **5.** 8 **6.** $\frac{1}{2}$

7·4 Mass and Weight

p. 315 **1.** 2 **2.** 7,000 **3.** 80 **4.** 2,500

7•5 Time

7•6 Size and Scale

Chapter 8 Tools

8•1 Four-Function Calculator

8•2 Scientific Calculator

8•3 Geometry Tools

8·4 Spreadsheets

Index

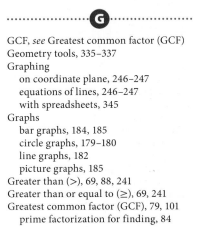

INDEX

Photo Credits

All coins photographed by United States Mint.
002–003 Seth Goldfarb/Getty Images; **064–065** Kris Timken/ Getty Images; **074** Peter Chadwick/Getty Images; **145** Getty Images; **176** PunchStock; **185** (l)Wil Meinderts/Getty Images, (c)GK Hart & Vikki Hart/Getty Images, (r)Colin Keates/Getty Images; **193** CORBIS; **198** Michael Rosenfeld/Getty Images; **217** Andy Crawford/Getty Images; **248** Getty Images; **259** CORBIS; **284** Doug Menuez/Getty Images; **297** Don Farrall/ Getty Images; **310** Andrew Ward/Getty Images; **318** Chris Johnson/Alamy; **330** Alamy; **350–351** Carson Ganci/CORBIS.